The **BEST WRITING** on **MATHEMATICS**

2021

The **BEST** **WRITING** on **MATHEMATICS**

Mircea Pitici, Editor

PRINCETON UNIVERSITY PRESS
PRINCETON AND OXFORD

Published by Princeton University Press
41 William Street, Princeton, New Jersey 08540
99 Banbury Road, Oxford OX2 6JX

press.princeton.edu

All Rights Reserved

ISBN 978-0-691-22571-5
ISBN (pbk.) 978-0-691-22570-8
ISBN (ebook) 978-0-691-22572-2

British Library Cataloging-in-Publication Data is available

Editorial: Susannah Shoemaker, Diana Gillooly and Kristen Hop
Production Editorial: Nathan Carr
Text Design: Carmina Alvarez
Cover Design: Chris Ferrante
Production: Jacqueline Poirier
Publicity: Matthew Taylor and Carmen Jimenez
Copyeditor: Paula Bérard

This book has been composed in Perpetua

Printed on acid-free paper. ∞

Printed in the United States of America

1 3 5 7 9 10 8 6 4 2

for Martha Hardesty

Contents

Color insert follows page 240

Introduction

MIRCEA PITICI

The Best Writing on Mathematics 2021 is the twelfth anthology in an annual series bringing together diverse perspectives on mathematics, its applications, and their interpretation—as well as on their social, historical, philosophical, educational, and interdisciplinary contexts. The volume should be seen as a continuation of the previous volumes. Since the series faces an uncertain future, I summarize briefly here its scorecard. We included 293 articles or book chapters in this series, written by almost 400 authors (several authors were represented in the series multiple times), as follows:

BWM Volume	Number of Pieces	Number of Authors
2010	36	43
2011	27	32
2012	25	29
2013	21	23
2014	24	33
2015	29	53
2016	30	41
2017	19	24
2018	18	33
2019	18	32
2020	20	23
2021	26	34
Totals	293	400

The pieces offered this time originally appeared during 2020 in professional publications and/or in online sources. The content of the volume is the result of a subjective selection process that started with many more candidate articles. I encourage you to explore the pieces that did not make it between the covers of this book; they are listed in the section of notable writings.

This introduction is shorter than the introduction to any of the preceding volumes. I made it a habit to direct the reader to other books on mathematics published recently; this time I will omit that part due to the unprecedented times we lived last year. The libraries accessible to me were closed for much of the research period I dedicated to this volume, and the services for borrowing physical books suffered serious disruptions. A few authors and publishers sent me volumes; yet mentioning here just those titles would be unfair to the many authors whose books I could not obtain.

Overview of the Volume

Once again, this anthology contains an eclectic mix of writings on mathematics, with a few even alluding to the events that just changed our lives in major ways.

To start, Viktor Blåsjö takes a cue from our present circumstances and reviews historical episodes of remarkable mathematical work done in confinement, mostly during wars and in imprisonment.

Andrew Lewis-Pye explains the basic algorithmic rules and computational procedures underlying cryptocurrencies and other blockchain applications, then discusses possible future developments that can make these instruments widely accepted.

Michael Duddy points out that the ascendancy of computational design in architecture leads to an inevitable clash between logic, intellect, and truth on one side—and intuition, feeling, and beauty on the other side. He explains that this trend pushes the decisions traditionally made by the human architect out of the resolutions demanded by the inherent geometry of architecture.

Steve Pomerantz combines elements of basic complex function mapping to reproduce marble mosaic patterns built during the Roman Renaissance of the twelfth and thirteenth centuries.

Ben Logsdon, Anya Michaelsen, and Ralph Morrison construct equations in two variables that represent, in algebraic form, geometric renderings of alphabet letters—thus making it possible to generate word-like figures, successions of words, and even full sentences through algebraic equations.

Maria Trnkova elaborates on crocheting as a medium for building models in hyperbolic geometry and uses it to find results of mathematical interest.

Yelda Nasifoglu decodes the political substrates of an anonymous seventeenth century play allegorically performed by geometric shapes.

In the next piece, Stephen K. Lucas, Evelyn Sander, and Laura Taalman present two methods for generating three-dimensional objects, show how these methods can be used to print models useful in teaching multivariable calculus, and sketch new directions pointing toward applications to dynamical systems.

Joshua Sokol tells the story of a quest to classify geological shapes mathematically—and how the long-lasting collaboration of a mathematician with a geologist led to the persuasive argument that, statistically, the most common shape encountered in the structure of the (under) ground is cube-like.

Don Monroe describes the perfect similarity between foundational algorithms in quantum computing and an experimental method for approximating the constant π, then asks whether it is indicative of a deeper connection between phenomena in physics and mathematics or it is a mere (yet striking) coincidence.

Kevin Hartnett relates recent developments in computer science and their unforeseen consequences for physics and mathematics. He explains that the equivalence of two classes of problems that arise in computation, recently proved, answers in the negative two long-standing conjectures: one in physics, on the causality of distant-particle entanglement, the other in mathematics, on the limit approximation of matrices of infinite dimension with finite-dimension matrices.

David Hand reviews the risks, distortions, and misinterpretations caused by missing data, by ignoring existing accurate information, or by falling for deliberately altered information and/or data.

In the same vein, Michael Wallace discusses the insidious perils introduced in experimental and statistical analyses by measurement errors and argues that the assumption of accuracy in the data collected from observations must be recognized and questioned.

In the midst of our book—like a big jolt on a slightly bumpy road—John Conway, Mike Paterson, and their fictive co-author Moscow, bring inimitable playfulness, characteristic brilliance, multiple puns, and nonexistent self-references to bear on an easy game of numbers that (dis)proves to be trickier than it seems!

Next, Sanjoy Mahajan explains (and illustrates with examples) why some mathematical formulas and some physical phenomena change expression at certain singular points.

Stan Wagon describes the counterintuitive movement of a bicycle pedal relative to the ground, also known as "the bicycle paradox," and uses basic trigonometry to elucidate the mathematics underlying the puzzle.

Jacob Siehler combines modular arithmetic and the theory of linear systems to solve a pyramid-coloring challenge.

Natalie Wolchover untangles threads that connect foundational aspects of numbers with logic, information, and physical laws.

The late Harold Edwards pleads for a reading of the classics of mathematics on their own terms, not in the altered "Whig" interpretation given to them by the historians of mathematics.

Michael Barany uncovers archival materials surrounding the birth circumstances, the growing pains, and the political dilemmas of the *Notices of the American Mathematical Society*—a publication initially meant to facilitate internal communication among the members of the world's foremost mathematical society.

Mike Askew pleads for raising reasoning in mathematics education at least to the same importance given to procedural competence—and describes the various kinds of reasoning involved in the teaching and learning of mathematics.

Roger Howe compares the professional opportunities for improvement and the career structure of mathematics teachers in China and in the United States—and finds that in many respects the Chinese ways are superior to the American practices.

Stephan Ramon Garcia draws on his work experience with senior undergraduate students engaged in year-end projects to distill two dozen points of advice for instructors who supervise mathematics research done by undergraduates.

Adam Glesser, Bogdan Suceavă, and Mihaela B. Vâjiac read (and copiously quote) Sophie Germain's French *Essays* (not yet translated into English) to unveil a mind not only brilliant in original mathematical contributions that stand through time, but also insightful in humanistic vision.

Melvyn Nathanson raises the puzzling issues of authorship, copyright, and secrecy in mathematics research, together with many related ethical and practical questions; he comes down uncompromisingly on the side of maximum openness in sharing ideas.

In the end piece of the volume, Terence Tao candidly recalls selected adventures and misadventures of growing into one of the world's foremost mathematicians.

⚭

This year has been difficult for all of us; each of us has been affected in one way or another by the current (as of May 2021) health crisis, some tragically. The authors represented in this anthology are no exception. For the first time since the series started, contributors to a volume passed away while the book was in preparation—in this case, John H. Conway (deceased from coronavirus complications) and Harold M. Edwards.

⚭

I hope you will enjoy reading this anthology at least as much as I did while working on it. I encourage you to send comments to Mircea Pitici, P.O. Box 4671, Ithaca, NY 14852; or electronic correspondence to mip7@cornell.edu.

Lockdown Mathematics:
A Historical Perspective

VIKTOR BLÅSJÖ

Isolation and Productivity

"A mathematician is comparatively well suited to be in prison." That was the opinion of Sophus Lie, who was incarcerated for a month in 1870. He was 27 at the time. Being locked up did not hamper his research on what was to become Lie groups. "While I was sitting for a month in prison . . ., I had there the best serenity of thought for developing my discoveries," he later recalled [11, pp. 147, 258].

Seventy years later, André Weil was to have a very similar experience. The circumstances of their imprisonments—or perhaps the literary tropes of their retellings—are closely aligned. Having traveled to visit mathematical colleagues, both found themselves engrossed in thought abroad when a war broke out: Lie in France at the outbreak of the Franco-Prussian War, and Weil in Finland at the onset of World War II. They were both swiftly suspected of being spies, due to their strange habits as eccentric mathematicians who incessantly scribbled some sort of incomprehensible notes and wandered in nature without any credible purpose discernible to outsiders. Both were eventually cleared of suspicion upon the intervention of mathematical colleagues who could testify that their behavior was in character for a mathematician and that their mysterious notebooks were not secret ciphers [11, pp. 13–14, 146–147; 13, pp. 130–134].

Weil was deported back to France, where he was imprisoned for another few months for skirting his military duties. Like Lie, he had a productive time in prison. "My mathematics work is proceeding beyond my wildest hopes, and I am even a bit worried—if it's only in prison that I work so well, will I have to arrange to spend two or three months locked up every year?" "I'm hoping to have some more time here to finish in peace and quiet what I've started. I'm beginning to think that nothing is more conducive to the abstract sciences than prison." "My sister says that when I leave here I should become a monk, since this regime is so conducive to my work."

Weil tells of how colleagues even expressed envy of his prison research retreat. "Almost everyone whom I considered to be my friend wrote me at this time. If certain people failed me then, I was not displeased to discover the true value of their friendship. At the beginning of my time in [prison], the letters were mostly variations on the following theme: 'I know you well enough to have faith that you will endure this ordeal with dignity.' . . . But before long the tone changed. Two months later, Cartan was writing: 'We're not all lucky enough to sit and work undisturbed like you.'" And Cartan was not the only one: "My Hindu friend Vij[ayaraghavan] often used to say that if he spent six months or a year in prison he would most certainly be able to prove the Riemann hypothesis. This may have been true, but he never got the chance."

But Weil grew weary of isolation. He tried to find joy in the little things: "[In the prison yard,] if I crane my neck, I can make out the upper branches of some trees." "When their leaves started to come out in spring, I often recited to myself the lines of the *Gita*: '*Patram puspam phalam toyam* . . .' ('A leaf, a flower, a fruit, water, for a pure heart everything can be an offering')." Soon he was reporting in his letters that "My mathematical fevers have abated; my conscience tells me that, before I can go any further, it is incumbent upon me to work out the details of my proofs, something I find so deadly dull that, even though I spend several hours on it every day, I am hardly getting anywhere" [13, pp. 142–150].

Judging by these examples, then, it would seem that solitary confinement and a suspension of the distractions and obligations of daily life could be very conducive to mathematical productivity for a month or two, but could very well see diminishing returns if prolonged. Of course, it is debatable whether coronavirus lockdown is at all analogous to these gentleman prisons of yesteryear. When Bertrand Russell was imprisoned for a few months for pacifistic political actions in 1918, he too "found prison in many ways quite agreeable. . . . I read enormously; I wrote a book, *Introduction to Mathematical Philosophy*." But his diagnosis of the cause of this productivity is less relatable, or at least I have yet to hear any colleagues today exclaiming about present circumstances that "the holiday from responsibility is really delightful" [9, pp. 29–30, 32].

Mathematics Shaped by Confinement

"During World War II, Hans Freudenthal, as a Jew, was not allowed to work at the university; it was in those days that his interest in mathematics education at primary school level was sparked by 'playing school' with his children—an interest that was further fueled by conversations with his wife." This observation was made in a recent editorial in *Educational Studies in Mathematics* [1]—a leading journal founded by Hans Freudenthal. Coronavirus lockdown has put many mathematicians in a similar position today. Perhaps we should expect another surge in interest in school mathematics among professional mathematicians.

Freudenthal's contemporary Jakow Trachtenberg, a Jewish engineer, suffered far worse persecution, but likewise adapted his mathematical interests to his circumstances. Imprisoned in a Nazi concentration camp without access to even pen and paper, he developed a system of mental arithmetic. Trachtenberg survived the concentration camp and published his calculation methods in a successful book that has gone through many printings and has its adherents to this day [12].

Another Nazi camp was the birthplace of "spectral sequences and the theory of sheaves . . . by an artillery lieutenant named Jean Leray, during an internment lasting from July 1940 to May 1945." The circumstances of the confinement very much influenced the direction of this research: Leray "succeeded in hiding from the Germans the fact that he was a leading expert in fluid dynamics and mechanics. . . . He turned, instead, to algebraic topology, a field which he deemed unlikely to spawn war-like applications" [10, pp. 41–42].

An earlier case of imprisonment shaping the course of mathematics is Jean-Victor Poncelet's year and a half as a prisoner of war in Russia. Poncelet was part of Napoleon's failed military campaign of 1812 and was only able to return to France in 1814. During his time as a prisoner, he worked on geometry. Poncelet had received a first-rate education in mathematics at the École Polytechnique, and his role in the military was as a lieutenant in the engineering corps. In his Russian prison, he did not have access to any books, so he had to work out all the mathematics he knew from memory. Perhaps it is only because mathematics lends itself so well to being reconstructed in this way that Poncelet ended up becoming a mathematician; other scientific or engineering interests would have been harder to pursue in isolation without books. The absence of books for reference would also naturally lead to a desire to unify geometrical theory and derive many results from a few key principles in Poncelet's circumstances. This is a prominent theme in early nineteenth-century geometry overall; it was not only the imprisoned who had this idea. But it is another sense in which Poncelet could make a virtue out of necessity with the style of mathematics he was confined to during his imprisonment.

The same can be said for another characteristic of early nineteenth-century geometry, namely, the prominent role of visual and spatial intuition. This too was a movement that did not start with Poncelet, but was fortuitously suited to his circumstances. Consider, for instance, the following example from the *Géométrie descriptive* of Monge, who had been one of Poncelet's teachers at the École Polytechnique. Monge was led to consider the problem of representing three-dimensional objects on a plane for purposes of engineering, but he quickly realized that such ideas can yield great insights in pure geometry as well, for instance, in the theory of poles and polars, which is a way of realizing the projective duality of points and lines. The foundation of this theory is to establish a bijection between the set of all points and the set of all lines in a plane. Polar reciprocation with respect to a circle associates a line with every point and a point with every line as follows. Consider a line that cuts through the circle (Figure 1). It meets the circle at two points. Draw the tangents to the circle through these points. The two tangents meet in a point. This point is the pole of the line. Conversely, the line is the polar of the point.

But what about a line outside the circle (or, equivalently, a point inside the circle, Figure 2)? Let *L* be such a line. For every point on *L*

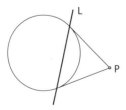

FIGURE 1. Polar reciprocation with respect to a circle: simplest case. Points *P* outside the circle are put in one-to-one correspondence with lines *L* intersecting the circle.

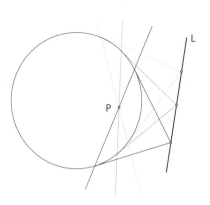

FIGURE 2. Polar reciprocation with respect to a circle: trickier case. Points *P* inside the circle are put in one-to-one correspondence with lines *L* that don't intersect the circle. The mapping works because of the collinearity of the meeting points of the tangents: a nontrivial result that becomes intuitively evident by introducing the third dimension and viewing the figure as the cross section of a configuration of cones tangent to a sphere.

there is a polar line through the circle, as above. We claim that all these polar lines have one point in common, so that this point is the natural pole of *L*. Monge proves this by cleverly bringing in the third dimension. Imagine a sphere that has the circle as its equator. Every point on *L* is the vertex of a tangent cone to this sphere. The two tangents to the equator are part of this cone, and the polar line is the perpendicular projection of the circle of intersection of the sphere and the cone. Now consider a plane through *L* tangent to the sphere. It touches the sphere at one point *P*. Every cone contains this point (because the line from any point on *L* to *P* is a tangent to the sphere and so is part of the tangent cone). Thus, for every cone, the perpendicular projection of the intersection with the sphere goes through the point perpendicularly below *P*, and this is the pole of *L*, and *L* is the polar of this point. *QED*

One is tempted to imagine that Poncelet was forced to turn to this intuitive style of geometry due to being deprived of pen and paper, just as Trachtenberg had to resort to mental arithmetic. But this is a half-truth at best, for Poncelet evidently did have crude writing implements at his disposal: the prisoners were allocated a minimal allowance, for which he was able to obtain some sheets of paper, and he also managed to make his own ink for writing [5, p. 20].

Ibn al-Haytham is another example of a mathematician starting out as an engineer and then turning increasingly to mathematics while in confinement. Early in his career, he devised an irrigation scheme that would harness the Nile to water nearby fields. When his plans proved unworkable, "he feigned madness in order to escape the wrath of the Caliph and was confined to a private house for long years until the death of the tyrannical and cruel ruler. He earned his livelihood by copying in secret translations of Euclid's and Ptolemy's works" [7, p. 156]. Euclidean geometry and Ptolemaic astronomical calculations are certainly better suited to house arrest scholarship than engineering projects. One may further wonder whether it is a coincidence that Ibn al-Haytham, who was forced to spend so many sunny days indoors, also discovered the camera obscura and gave it a central role in his optics.

From these examples, we can conclude that if coronavirus measures are set to have an indirect impact on the direction of mathematical research, it would not be the first time lockdown conditions have made one area or style of mathematics more viable than another.

Newton and the Plague

Isaac Newton went into home isolation in 1665, when Cambridge University advised "all Fellows & Scholars" to "go into the Country upon occasion of the Pestilence," since it had "pleased Almighty God in his just severity to visit this towne of Cambridge with the plague" [14, p. 141]. Newton was then 22 and had just obtained his bachelor's degree. His productivity during plague isolation is legendary: this was his *annus mirabilis*, marvelous year, during which he made a number of seminal discoveries. Many have recently pointed to this as a parable for our time, including, for instance, the *Washington Post* [3]. The timeline is none too encouraging for us to contemplate: the university effectively remained closed for nearly two years, with an aborted attempt at reopening halfway through, which only caused "the pestilence" to resurge.

It is true that Newton achieved great things during the plague years, but it is highly doubtful whether the isolation had much to do with it, or whether those years were really all that much more *mirabili* than others. Newton was already making dramatic progress before the plague broke out and was on a trajectory to great discoveries regardless of public health regulations. Indeed, Newton's own account of how much he accomplished "in the two plague years of 1665 & 1666" attributes his breakthroughs not to external circumstances but to his inherent intellectual development: "For in those days I was in the prime of my age for invention & minded Mathematicks & Philosophy more then at any time since" [15, p. 32].

"Philosophy" here means physics. And indeed, in this subject Newton did much groundwork for his later success during the plague years, but the fundamental vision and synthesis that we associate with Newtonian mechanics today was still distinctly lacking. His eventual breakthrough in physics depended on interactions with colleagues rather than isolation. In 1679, Hooke wrote to Newton for help with the mathematical aspects of his hypothesis "of compounding the celestiall motions of the planetts of a direct motion by the tangent & an attractive motion towards the centrall body." At this time, "Newton was still mired in very confusing older notions." To get Newton going, Hooke had to explicitly suggest the inverse square law and plead that "I doubt not but that by your excellent method you will easily find out what that Curve [the orbit] must be." Only then, "Newton quickly broke through to dynamical enlightenment . . . following [Hooke's] signposted track" [2, pp. 35–37, 117].

Newton later made every effort to minimize the significance of Hooke's role. Indeed, Hooke was just one of many colleagues who ended up on Newton's enemies list. This is another reason why Newton's plague experience is a dubious model to follow. Newton could be a misanthropic recluse even in normal times. When Cambridge was back in full swing, Newton still "seldom left his chamber," contemporaries recalled, except when obligated to lecture—and even that he might as well have done in his chamber for "ofttimes he did in a manner, for want of hearers, read to the walls" [4, n. 11]. He published reluctantly, and when he did, Newton "was unprepared for anything except immediate acceptance of his theory": "a modicum of criticism sufficed, first to incite him to rage, and then to drive him into isolation" [14, pp. 239, 252]. With Hooke, as with so many others, it may well be that Newton only ever begrudgingly interacted with him in the first place for the purpose of proving his own superiority. But that's a social influence all the same. Even if Hooke's role was merely to provoke a sleeping giant, the fact remains that Newton's *Principia* was born then and not in quarantine seclusion.

In mathematics, it is accurate enough to say that Newton "invented calculus" during the plague years. But he was off to a good start already before then, including the discovery of the binomial series. In optics,

Newton himself said that the plague caused a two-year interruption in his experiments on color that he had started while still at Cambridge [6, p. 31]. Perhaps this is another example of pure mathematics being favored in isolation at the expense of other subjects that are more dependent on books and tools.

Home isolation also affords time for extensive hand calculations: a self-reliant mode of mathematics that can be pursued without library and laboratory. Newton did not miss this opportunity during his isolation. As he later recalled, "[before leaving Cambridge] I found the method of Infinite series. And in summer 1665 being forced from Cambridge by the Plague I computed ye area of ye Hyperbola . . . to two & fifty figures by the same method" [14, p. 98]. Newton's notebook containing this tedious calculation of the area under a hyperbola to 52 decimals can be viewed at the Cambridge University Library website [8].

References

1   Arthur Bakker and David Wagner, Pandemic: Lessons for today and tomorrow? *Educational Studies in Mathematics*, 104 (2020), 1–4, https://doi.org/10.1007/s10649-020-09946-3.

2   Zev Bechler, Ed., *Contemporary Newtonian Research*, Reidel, Dordrecht, Netherlands, 1982.

3   Gillian Brockell, During a pandemic, Isaac Newton had to work from home, too. He used the time wisely, *Washington Post*, March 12, 2020.

4   I. Bernard Cohen, *Newton, Isaac,* Dictionary of Scientific Biography, Vol. 10, Charles Scribner's Sons, New York, 1974.

5   Isidore Didion, *Notice sur la vie et les ouvrages du Général J.-V. Poncelet*, Gauthier-Villars, Paris, 1869.

6   A. Rupert Hall, *Isaac Newton: Adventurer in Thought*, Cambridge University Press, Cambridge, U.K., 1992.

7   Max Meyerhof, Ali al-Bayhaqi's Tatimmat Siwan al-Hikma: A biographical work on learned men of the Islam, *Osiris* 8 (1948), 122–217.

8   Isaac Newton, MS Add. 3958, 79r ff., https://cudl.lib.cam.ac.uk/view/MS-ADD-03958/151.

9   Bertrand Russell, *The Autobiography of Bertrand Russell: 1914–1944*, Little Brown and Company, Boston, 1968.

10 Anna Maria Sigmund, Peter Michor, and Karl Sigmund, Leray in Edelbach. *Mathematical Intelligencer* 27 (2005), 41–50.

11 Arild Stubhaug, *The Mathematician Sophus Lie: It Was the Audacity of My Thinking*, Springer, Berlin, 2002.

12 Jakow Trachtenberg, *The Trachtenberg Speed System of Basic Mathematics*, Doubleday and Company, New York, 1960.

13 André Weil, *The Apprenticeship of a Mathematician*, Birkhäuser, Basel, Switzerland, 1992.

14 Richard S. Westfall, *Never at Rest: A Biography of Isaac Newton*, Cambridge University Press, Cambridge, U.K., 1983.

15 D. T. Whiteside, Newton's Marvellous Year: 1666 and All That, *Notes and Records of the Royal Society of London* 21(1) (1966), 32–41.

Cryptocurrencies: Protocols for Consensus

ANDREW LEWIS-PYE

The novel feature of Bitcoin [N+08] as a currency is that it is designed to be *decentralized*, i.e., to be run without the use of a central bank, or any centralized point of control. Beyond simply serving as currencies, however, cryptocurrencies like Bitcoin are really protocols for reaching consensus over a decentralized network of users. While running currencies is one possible application of such protocols, one might consider broad swaths of other possible applications. As one example, we have already seen cryptocurrencies used to instantiate *decentralized autonomous organizations* [KOH19], whereby groups of investors come together and coordinate their investments in a decentralized fashion, according to the rules of a protocol that is defined and executed "on the blockchain." One might also envisage new forms of decentralized financial markets, or perhaps even a truly decentralized World Wide Web, in which open-source applications are executed by a community of users, so as to ensure that no single entity (such as Google or Facebook) exerts excessive control over the flow of personal data and other information.

Many questions must be answered before we can talk with any certainty about the extent to which such possibilities can be realized. Some of these questions concern human responses, making the answers especially hard to predict. How much appetite does society have for decentralized applications, and (beyond the possibilities listed above) what might they be? In what contexts will people feel that the supposed advantages of decentralization are worth the corresponding trade-offs in efficiency? There are also basic technical questions to be addressed. Perhaps the best known of these is the so-called *scalability* issue: Can cryptocurrency protocols be made to handle transactions at a rate sufficient to make them useful on a large scale?

In this paper, we will describe how Bitcoin works in simple terms. In particular, this means describing how the Bitcoin protocol uses hard computational puzzles in order to establish consensus as to who owns what. Then we will discuss some of the most significant technical obstacles to the large-scale application of cryptocurrency protocols and approaches that are being developed to solve these problems.

Bitcoin and Nakamoto Consensus

The Bitcoin network launched in January 2009. Since that time, the total value of the currency has been subject to wild fluctuations, but at the time of writing, it is in excess of $170 billion.[1] Given the amount of attention received by Bitcoin, it might be surprising to find out that consensus protocols have been extensively studied in the field of distributed computing since at least the 1970s [Lyn96]. What differentiates Bitcoin from previous protocols, however, is the fact that it is a *permissionless* consensus protocol; i.e., it is designed to establish consensus over a network of users that anybody can join, with as many identities as they like in any role. Anybody with access to basic computational hardware can join the Bitcoin network, and users are often *encouraged* to establish multiple public identities, so that it is harder to trace who is trading with whom.

It is not difficult to see how the requirement for permissionless entry complicates the process of establishing consensus. In the protocols that are traditionally studied in distributed computing, one assumes a fixed set of users, and protocols typically give performance guarantees under the condition that only a certain minority of users behave improperly— "improper" action might include malicious action by users determined to undermine the process. In the permissionless setting, however, users can establish as many identities as they like. Executing a protocol that is only guaranteed to perform well when malicious users are in the minority is thus akin to running an election in which people are allowed to choose their own number of votes.

In the permissionless setting, one therefore needs a mechanism for weighting the contribution of users that goes beyond the system of "one user, one vote." The path taken by Bitcoin is to weight users according to their computational power. This works because computational power is a scarce and testable resource. Users might be able to double their number of identities in the system at essentially zero cost, but this will not impact their level of influence. To do that, they will need to increase their computational power, which will be expensive.

Before we see how Bitcoin achieves this in more detail, we will need to get a clearer picture of how one might go about running a (centralized or decentralized) digital currency in the first place. To explain that, we will need some basic tools from cryptography.

BASIC TOOLS FROM CRYPTOGRAPHY

The two basic tools that we will need from cryptography are *signature schemes* and *hash functions*. Luckily, we can entirely black-box the way in which these tools are implemented. All that is required now is to understand the functionality that they provide.

SIGNATURE SCHEMES. Presently, most cryptocurrencies use signature schemes that are implemented using elliptic curve cryptography. The functionality provided by these signature schemes is simple. When one user wishes to send a message to another, the signature scheme produces a *signature*, which is specific to that message and that user. This works in such a way that any user receiving the message together with the signature can efficiently verify from whom the message came. So the use of an appropriate signature scheme means one cannot produce "fake" messages purporting to be from other users.

HASH FUNCTIONS. Hash functions take binary strings of any length as input and produce strings of a fixed length as output. Normally, we work with hash functions that produce 256-bit strings, and the 256-bit output is referred to as the *hash* of the input string. Beyond that basic condition, a hash function is designed to be as close as possible to being a random string generator (subject to the condition that the same input always gives the same output): Informally speaking, the closer to being a random string generator, the better the hash function. This means that a good hash function will satisfy two basic properties:

(a) Although in theory the function is not injective, in practice we will never find two strings that hash to the same value, because there are 2^{256} possible outputs.

(b) If tasked with finding a string that hashes to a value with certain properties, there is no more efficient method than trying inputs to the hash function one at a time, and seeing what they produce.

So if we are working with a good hash function, and we are tasked with finding a string of a certain length that hashes to a value starting with 10 zeros, then there is no more fundamentally efficient method than just plugging in input values until we find one that works.

Implementing a Centralized Digital Currency

As we have already said, it is the aim of Bitcoin to be decentralized. To understand what difficulties we face in implementing a decentralized digital currency, however, it is instructive to consider first how one might implement a *centralized* digital currency, which works *with* the use of a central bank. Once that simple case is dealt with, we can properly analyze what difficulties arise in the decentralized case.

Presumably, we want our currency to be divided into units, or *coins*. For the sake of simplicity, we will start by concentrating on what happens to a single coin and suppose that this coin is indivisible. So the owner of the coin can either spend the whole coin or nothing—they are not allowed to spend half the coin. In that case, we might have the "coin" simply be a *ledger* (i.e., an accounting record), which records its sequence of owners. A coin is thus a binary sequence, which could be visualized as below.

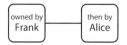

For now, do not not worry about how Frank came to own this coin in the first place—we will come to that later. Instead, let us consider what needs to happen when Alice, who presently owns the coin, wants to transfer it to another user. In the presence of a central bank, this is simple: Alice can form a new version of the coin, recording that it now belongs to the new user, Bob, say, and send that new version of the coin to the central bank. In order that the central bank can be sure that the extension to the ledger really was created by Alice, though, she will need to add her signature—we will picture the relevant signature as a little black box attached to the bottom right corner of that part of the ledger. Of course, if Alice has to add her signature now, Frank will also have had to add his signature when he transferred the coin to Alice. The signature added to each extension of the ledger can be seen as testimony by the previous owner that they wish to transfer the coin to the new user. The new version of the coin can then be represented as below.

When the central bank sees the new version of the coin, they can check to see that the signature is correct, and, if so, record the transaction as *confirmed*. The use of a signature scheme therefore suffices to ensure that only Alice can spend her coin. This is not the only thing we have to be careful about, though. We also need to be sure that Alice cannot spend her coin twice. In the presence of the central bank, this is also simple. Suppose Alice later creates a new version of the coin, which transfers the coin to another user, Charlie, instead. In this case, the central bank will see that this transaction conflicts with the earlier one that they have seen and so will reject it.

This simple protocol therefore achieves two basic aims:

1. Only Alice can spend her coin, and
2. Alice cannot "double spend."

So what changes when we try to do without the use of a central bank? Let us suppose that all users now store a copy of the coin. When Alice wishes to transfer the coin to Bob, she forms a new version of the coin, together with her signature, as before. Now, however, rather than sending it to the central bank, she simply sends the new version to various people in the network of users, who check the signature and then distribute it on to others, and so on. In this case, the use of signatures still suffices to ensure that only Alice can spend her coin. The issue is now that it becomes tricky to ensure that Alice cannot spend her coin twice. Alice could form two new versions of the coin, corresponding to two different transactions. If we could be certain that all other users saw these two versions in the same order, then there would not be a problem, as then users could just agree not to allow the second transaction. Unfortunately, we have no way of ensuring this is the case.[2]

Removing the Central Bank

From the discussion above, it is clear that we need a protocol for establishing irreversible consensus on transaction ordering. To describe how this can be achieved, we will initially describe a protocol that differs from Bitcoin in certain ways, and then we will describe what changes are required to make it the same as Bitcoin later.

Previously, we simplified things by concentrating on one coin. Let us now drop that simplification, and have all users store a *universal ledger*, which records what happens to all coins. We can also drop the simplification that coins are indivisible if we want, and allow transactions which transfer partial units of currency. So according to this modified picture, each user stores a universal ledger, which is just a "chain" of signed transactions. Each transaction in this chain might now follow an unrelated transaction, which transfers a different coin (or part of it) between a different pair of users: The universal ledger is just a chain of transactions recording all transfers of currency that occur between users.

The reader will notice that in the picture above, we have each transaction *pointing to* the previous transaction. We should be clear about how this is achieved, because it is important that we create a tamper-proof ledger: We do not want a malicious user to be able to remove intermediate transactions and produce a version of the universal ledger that looks valid. What we do is to have each signed transaction include the hash of the previous transaction as part of its data. Since hash values are (in effect) unique, this hash value serves as a unique identifier.

What happens next is the key new idea:

(A) We specify a computational puzzle corresponding to each transaction, which is specific to the transaction, and which can be solved only with a lot of computational work. The puzzle is chosen so that, while the solution takes a lot of computational work to find, a correct solution can easily (i.e., efficiently) be verified as correct. The solution to the puzzle corresponding to a given transaction is called a "proof of work" (PoW) for that transaction.

(B) We insist that a transaction cannot be included in the universal ledger unless accompanied by the corresponding PoW.

Do not worry immediately about precisely how the PoW is specified—we will come back to that shortly. Now when Alice wants to spend her coin, she sends the signed transaction out into the network of users, all of whom start trying to produce the necessary PoW. Only once the PoW is found can the transaction be appended to the universal ledger. So now transactions are added to the chain at the rate at which PoWs are found by the network of users. The PoWs are deliberately constructed to require time and resources to complete. Exactly how difficult they are to find is the determining factor in how fast the chain grows.

Of course, the danger we are concerned with is that a malicious user might try to form alternative versions of the ledger. How are we to know which version of the ledger is "correct"? In order to deal with these issues, we make two further stipulations (the way in which these stipulations prevent double spending will be explained shortly):

(C) We specify that the "correct" version of the ledger is the longest one. So when users create new transactions, they are asked to have these extend the "longest chain" of transactions (with the corresponding PoWs supplied) they have seen.

(D) For a certain *security parameter* k, a given user will consider a transaction t as "confirmed" if t belongs to a chain C which is at least k transactions longer than any they have seen that does not include t, and if t is followed by at least k many transactions in C.

The choice of k will depend on how sure one needs to be that double spending does not occur. For the sake of concreteness, the reader might think of $k = 6$ as a reasonable choice.

These are quite simple modifications. How do they prevent double spending? The basic idea is as follows. Suppose that at a certain point in time, Alice wants to double spend. Let us suppose that the longest chain of transactions is as depicted below, and that the confirmed transaction t that Alice wants to reverse is the third one (circled).

In order to reverse this transaction, Alice will have to form a new chain that does not include *t*. This means branching off before *t*, and building from there.

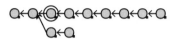

For people to believe it, however, this new chain will have to be the longest chain. The difficulty for Alice is that while she builds her new chain of transactions, the *rest of the network combined* is working to build the other longer chain.

So long as Alice does not have more computational power than the rest of the network combined, she will not be able to produce PoWs faster than they can. Her chain will therefore grow at a slower rate than the longest chain, and her attempt to double spend will fail.[3] So long as no malicious user (or coordinated set of users) has more power than the rest of the network combined, what we have achieved is a tamper-proof universal ledger, which establishes irreversible consensus on transaction ordering, and which operates in a decentralized way.

To finish this section, we now fulfill some earlier promises. We have to explain how PoWs are defined, what changes are necessary to make the protocol like Bitcoin, and how users come to own coins in the first place.

Defining PoWs

In fact, it will be useful to define PoWs for binary strings more generally—of course, transactions are specified by binary strings of a particular sort. To do this, we fix a good hash function *h*, and work with a difficulty parameter *d*, which (is not to be confused with the security parameter *k* and) can be adjusted to determine how hard the PoW is to find. For two strings *x* and *y*, let *xy* denote the concatenation of *x* and *y*. Then we define a PoW for *x* to be any string *y* such that $h(xy)$ starts with *d* many zeros. Given the properties of a good hash function described earlier, this means that there is no more efficient way to find a PoW for *x* than to plug through possible values for *y*, requiring 2^d

many attempts on average. The expected time it will take a user to find a PoW is therefore proportional to the rate at which they can process hash values, and for larger d, the PoW will be harder to find. Defining PoW in this way also means that the process by which the network as a whole finds PoWs can reasonably be modeled as a Poisson process: In any second, there is some independent probability that a PoW will be found, and that probability depends on the rate at which the network as a whole can process hashes.

USING BLOCKS OF TRANSACTIONS

The most significant difference between the protocol we have described and Bitcoin is that in Bitcoin the ledger does not consist of individual transactions, but *blocks* of transactions (hence the term "blockchain"). Each block is a binary string, which contains within its data a few thousand transactions,[4] together with a hash value specifying the previous block. So now, individual transactions are sent out into the network, as before. Rather than requiring a PoW for each individual transaction, however, Bitcoin asks users to collect large sets of transactions into blocks and only requires one PoW per block. The main reason[5] for this is worth understanding properly, because it also relates quite directly to the issue of scalability, which we will discuss in the next section. The key realization here is that we have to take careful account of the fact that the underlying communication network has *latency*; i.e., it takes time for messages to propagate through the network. This latency becomes especially problematic when we work at the level of individual transactions, since they are likely to be produced at a rate that is high compared to network latency. For the sake of concreteness, it may be useful to work with some precise numbers. So, as an example, let us suppose that it takes 10 seconds for a transaction to propagate through the network of users. Suppose that we are using the protocol as defined previously, so that PoWs are required for individual transactions, rather than for blocks. To begin with, let us suppose that the difficulty parameter is set so that the network as a whole finds PoWs for transactions once every 10 minutes on average. Consider a point in time at which all users have seen the same longest chain C, and consider what happens when a PoW for a new transaction t_1 is found by a certain user, so that t_1 can be appended to C. The PoW for t_1 then begins to propagate through the network. The crucial observation is that there is then the following danger: During those 10 seconds of propagation time, there is some chance that another user, who has not yet seen the new extended version of the ledger, will find a PoW for another transaction t_2. In this case, we now have an honestly produced *fork*, which splits the honest users.

While some users will be looking to find PoWs to extend one version of the chain, others will be working to extend the other. At least briefly, this makes it slightly easier for a malicious user to double spend, because now they only have to outcompete each component of the divided network.

In that example, the chance of a fork is quite low, because PoWs are produced only once every 10 minutes on average, while propagation time is 10 seconds. If we have a PoW produced every minute on average, however, then the appearance of a fork will now be 10 times as likely. The problem we have is that, practically speaking, we will need transactions to be processed at a much higher rate than one per minute: Bitcoin can process 7 transactions per second, and this is generally regarded as being unacceptably slow for large-scale adoption. If PoWs are being produced at a rate of 7 per second, then we will not only see forks of the kind described above. We will see forks within forks within forks, with different honest users split between many different chains, and the security of the protocol will be dramatically compromised. This problem is avoided by using blocks, because blocks of transactions can be produced much more slowly: In Bitcoin, the difficulty of the PoW is adjusted so as to ensure that one block is produced every 10 minutes on average. This means that, most of the time, all honest users will be working to extend the same chain.

MINTING NEW COINS

In Bitcoin, the users who look to provide the PoW for blocks of transactions are referred to as "miners," and the process of searching for PoWs is called "mining." Now, though, we have a problem of incentives to deal with. Mining costs money. There are hardware and electricity costs, among others. If the system is to be secure against double spending, then we certainly need lots of money to be spent on mining; the security of the system is directly determined by how much it would cost a malicious user to establish more mining power than the rest of the network. To incentivize them to mine, this means that miners need to be paid in one way or another, and it is here that it is rather convenient that we happen to be designing a currency. In the context of running a currency, the solution becomes simple: We reward miners for finding PoWs by giving them currency.[6] This also, rather neatly, solves the problem as to how users come to own coins in the first place. It is when miners find a PoW that they are assigned previously unowned units of currency.

The Issue of Scalability

Various technical issues need to be addressed before cryptocurrencies see large-scale adoption. Among the less serious of these is that Bitcoin requires fully participating users to download the *entire* ledger (presently more than 200 GB). Much more significant is the fact that PoW protocols are energy intensive, to the point that recent estimates show Bitcoin consuming more energy than the nation of Switzerland.[7] The question that has received the most attention, however, is how to increase transaction rates: While Visa is capable of handling more than 65,000 transactions per second, Bitcoin can presently process 7 transactions per second. In this section, we explain why Bitcoin processes transactions so slowly, and some proposed solutions.

THE TWO TRANSACTION RATE BOTTLENECKS

There are two fundamental bottlenecks that limit transaction rates for cryptocurrency protocols such as Bitcoin.

THE LATENCY BOTTLENECK.   The Bitcoin protocol limits the size of blocks to include a few thousand transactions, and the PoW difficulty setting is adjusted every couple of weeks, so that blocks are produced once every 10 minutes on average. These two factors—the cap on the size of blocks and the fixed rate of block production—directly result in the limited transaction rate described above. To increase transactions rates, though, it is tempting to think that one could simply increase the size of blocks, or have them produced more frequently. To increase the transaction rate by a factor of 600, why not have blocks being produced once per second? In fact, the issue here is precisely the same as the motivation for using blocks in the first place, which we discussed in detail previously.[8] Our earlier discussion considered individual transactions, but precisely the same argument holds for blocks of transactions: The fact that the network has *latency* (blocks take a few seconds to propagate through the network) means that whenever a block is produced, there is also the possibility of an honestly produced fork in the blockchain. If we double the rate of block production, then we double the probability of that fork. If we were to have a block produced once per second on average, then we would see forks within forks within forks, and the protocol would no longer be secure.[9] Essentially the same analysis holds in the case that we increase the size of blocks, because doing so increases propagation time. This increase in propagation time similarly increases the probability of a fork.

THE PROCESSOR BOTTLENECK.   A basic feature of Bitcoin that distinguishes it from centrally run currencies is that all fully participating users are required to process all transactions. For some applications of blockchain technology, however, one might want to process many millions of transactions per second.[10] To achieve this (even if one solves the latency bottleneck), one needs to deal with the fundamental limitation that transactions can only be processed as fast as can be handled by the slowest user required to process all transactions. The prospect of a decentralized Web 3.0 in which all users have to process all interactions must surely be a nonstarter. So how can one work around this? Limiting the users who have to process all transactions to a small set with such capabilities constitutes a degree of centralization. Another possibility is not to require *any* users to process all transactions. For example, one might consider a process called "sharding," whereby one runs a large number of blockchains that allow limited interactions between them, while requiring each user individually to process transactions on a small set of blockchains at any given time.

SOLUTIONS IN THREE LAYERS

A multitude of mechanisms have been proposed with the aim of increasing transaction rates. They can be classified as belonging to three *layers*.

LAYER 0. These are solutions that do not involve modifying the protocol itself, but aim instead to improve on the underlying infrastructure used by the protocol. Layer 0 solutions range from simply building a faster Internet connection, to approaches such as bloXroute [KBKS18], a blockchain distribution network, which changes the way in which messages propagate through the network. At this point, Layer 0 solutions are generally best seen as approaches to dealing with the latency bottleneck.

LAYER 1. These solutions involve modifying the protocol itself, and they can be aimed at dealing with either the latency bottleneck or the processor bottleneck.

LAYER 2. These protocols are implemented *on top of* the underlying cyrptocurrency. So the underlying cryptocurrency is left unchanged, and one runs an extra protocol which makes use of the cryptocurrency's blockchain. Generally, the aim is to outsource work so that most transactions can take place "off-chain," with the underlying cryptocurrency blockchain used (hopefully rarely) to implement conflict resolution. To make these ideas more concrete, we later explain the basic idea behind the Lightning Network, which is probably the best known Layer 2 solution. Layer 2 solutions are generally aimed at solving the processor bottleneck.

To finish this section, we describe two well-known scalability solutions. Due to the limited available space, we do not say anything further about Layer 0 solutions. We briefly discuss a Layer 1 solution called the GHOST protocol [SZ15], which aims at dealing with the latency bottleneck. Then we explain the basic idea behind the Lightning Network [PD16], already mentioned above as a Layer 2 solution aimed at solving the processor bottleneck.

THE GHOST PROTOCOL

Recall that the latency bottleneck was caused by forks: While Bob is waiting for confirmation on a transaction in which Alice sends him money, a fork in the blockchain may split the honest users of the network. Suppose that the transaction is in the block B_1 in the picture below.

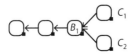

If the honest users are split between chains C_1 and C_2, then these will each grow more slowly than if there was a single chain. This makes it easier for Alice to form a longer chain.

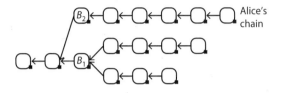

The solution proposed by the GHOST (Greedy Heaviest Observed SubTree) protocol is simple. Rather than selecting the longest chain, we select blocks according to their total number of descendants. This means selecting the chain inductively: Starting with the first block (the so-called "genesis" block), we choose between children by selecting that with the greatest total number of descendants, and then iterate this process to form a longer chain, until we come to a block with no children. This way B_1 will be selected over B_2 in the picture above, because B_1 has seven descendants, while B_2 only has five. So the consequence of using the GHOST protocol is that forks *after* B_1 do not matter, in the sense that they do not change the number of descendants of B_1, and so do not increase Alice's chance of double spending. We can increase the rate of block production, and although there will be an increase in the number of forks, Alice will still require more computational power than the rest of the network combined to double spend.

Unfortunately, however, this modified selection process gives only a partial solution to the latency bottleneck. The reason is that while forks *after* B_1 now do not matter (for confirmation of B_1), forks *before* B_1 still do. To see why, recall that, in order to be confirmed, B_1 must belong to a chain that is longer by some margin than any not including B_1.

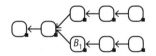

If blocks are produced at a rate that is low compared to the time it takes them to propagate through the network, then such (possibly honestly produced) ties are unlikely to persist for long—before too long, an interval of time in which no blocks are produced will suffice to break the tie. If the rate of block production is too fast, however, then such ties may extend over long periods. This produces long confirmation times.

In summary, the GHOST protocol allows us to increase the rate of block production without decreasing the proportion of the network's computational power that Alice will need to double spend. If we increase the rate too much, however, this will result in extended confirmation times.

THE LIGHTNING NETWORK

In order to explain the Lightning Network, we first need to discuss "smart contracts."

SMART CONTRACTS. So far, we have considered only very simple transactions, in which one user pays another in a straightforward fashion: Alice transfers funds to Bob, in such a way that Bob's signature now suffices to transfer the funds again. Bitcoin does allow, though, for more sophisticated forms of transaction. One might require two signatures to spend money, for example, or perhaps any two from a list of three signatures—so now units of currency might be regarded as having multiple "owners." In such a situation, where there are many forms a transaction could take, how is Alice to specify the transaction she wants to execute? The approach taken by Bitcoin is to use a "scripting language," which allows users to describe how a transaction should work. While Bitcoin has a fairly simple scripting language, other cryptocurrencies, such as Ethereum [W+14], use scripting languages that are sophisticated enough to be *Turing complete*—this means that transactions can be made to simulate any computation in any programming language. As a mathematically minded example, (in principle) one might publish a transaction to the blockchain that automatically pays one million units of currency to anybody who can produce a (suitably encoded) proof of the Riemann hypothesis![11] This is also a functionality whose significance depends on the information available to such computations: If reliable information on stock markets and cryptocurrency prices were to be recorded on the blockchain, then it would immediately become possible to simulate futures, options, and essentially any financial product that can be programmed using the given information. For our purposes now, the point is this: Transactions can be specified to work in much more sophisticated ways than simply transferring currency from one user to another.

A BIDIRECTIONAL PAYMENT CHANNEL. The aim of the Lightning Network is to allow most transactions to take place "off-chain." This is achieved by establishing an auxiliary network of "payment channels." Before coming to the network as a whole, let us consider briefly how to implement an individual channel between two users.[12]

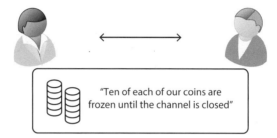

"Ten of each of our coins are frozen until the channel is closed"

So let us suppose that Alice and Bob wish to set up a payment channel between them. To initiate the channel, they will need to send one transaction to the underlying blockchain. This transaction is signed by both of them and says (in effect) that a certain amount of each of their assets should be frozen until the payment channel is "closed"—closing the channel has a precise meaning that we discuss shortly. For the sake of concreteness, let us suppose that they each freeze 10 coins. Once the channel is set up, Alice and Bob can now trade off-chain, simply by signing a sequence of time-stamped IOUs. If Alice buys something for three coins from Bob, then they both sign a time-stamped IOU stating that Alice owes Bob three coins. If Bob then buys something for one coin from Alice, they both sign a (later) time-stamped IOU stating that Alice now owes Bob two coins. They can continue in this way, so long as neither ever owes the other more than the 10 coins they have frozen. When either user wants to close the channel, they send in the most recent IOU to the blockchain, so that the frozen coins can be distributed to settle the IOU. We must guard against the possibility that the IOU sent is an old one, however. So, once an attempt is made to close the channel, we allow a fixed duration of time for the other user to counter with a more recent IOU.

THE NETWORK. The bidirectional payment channel described above required one transaction in the blockchain to set up, and a maximum of two to close. The system really becomes useful, however, once we have established an extensive network of payment channels.

Suppose now that Alice wishes to pay Derek, but that they have not yet established a payment channel. They *could* set up a new channel, but this would require sending transactions to the blockchain. Instead, Alice can pay Derek via Bob and Charlie, if those existing channels are already in place. Of course, we have to be careful to execute this so that no middleman can walk away with the money, but this can be achieved fairly simply, with the appropriate cryptographic protocols.

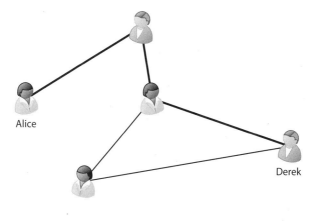

Discussion

Academically, the study of permissionless distributed computing protocols is in its early phases and is fertile territory for theoreticians, with much work to be done. Recent work [LPR20, GKL15, PSS17] has begun the process of establishing the same sort of framework for the rigorous analysis of permissionless protocols as was developed for permissioned protocols over many years. The hope is that, through the development of appropriate frameworks, a theory can be developed that probes the limits of what is possible through the development of impossibility results, as well as the formal analysis of existing protocols. Although the Bitcoin protocol was first described more than a decade ago, the original paper did not provide a rigorous security analysis. Since then, a number of researchers have done great work toward providing such an analysis [Ren19, GKL15, PSS17], but the development of appropriate frameworks for security analysis remains an ongoing task. In addressing the issue of scalability, and in dealing with the substantial issues of privacy and transparency that arise in connection with the use of cryptocurrencies, there is also plenty of scope for the use of more advanced cryptographic methods, such as succinct zero-knowledge proofs [BSBHR18].

Of course, there are many questions and issues that we have not had space to discuss. For example, it remains an ongoing task to develop a thorough *incentive*-based analysis of Bitcoin and other protocols: The protocol may behave well when only a minority of users (weighted by computational power) behave badly, but are the other "honest" users properly incentivized to follow the protocol? Is following the protocol a Nash equilibrium according to an appropriate set of payoffs? In fact, these questions have been shown to be somewhat problematic for Bitcoin. There are contexts in which miners are incentivized to deviate from the protocol [ES14], and the infrequent nature of miner rewards also means that miners are incentivized to form large "mining pools." Today, a small handful of mining pools carry out the majority of the mining for Bitcoin, meaning that control of the currency is really quite centralized.

Earlier, we briefly mentioned the significant issue that PoW protocols are energy intensive. A viable alternative to PoW may be provided by "proof-of-stake" (PoS): With a PoS protocol, users are selected to update state (i.e., to do things like publish blocks of transactions) with probability proportional to how much currency they own, rather than their computational power. PoS protocols face a different set of technical challenges [LPR20]. There are good reasons to believe, however, that as well as being energy efficient, PoS protocols may offer significant benefits in terms of increased security and decentralization.

At this point, it seems likely that substantial increases in transaction rates will be made possible over time through a combination of approaches. At least in the short to medium term, however, if we are to see large-scale adoption of cryptocurrencies, then one might conjecture that this is likely to be in applications such as the financial markets, where computational efficiency is important to a point, but where market efficiencies are key.

Notes

1. For an up-to-date value, see https://coinmarketcap.com.

2. Readers are encouraged to convince themselves that there is no simple solution here. For example, it might be tempting to think that one should cancel both if one sees contradictory transactions, but this will allow Alice to invalidate transactions deliberately after they are considered to have cleared.

3. A caveat is that finding a PoW is best modeled as probabilisitic, so there will be some chance that Alice will succeed in double spending, but it will be small.

4. At the time of writing, the monthly mean is just over 2,000 transactions per block.

5. There is a second reason. We want the rate at which PoWs are found, rather than the rate at which users wish to execute transactions, to be the determining factor in how fast the chain grows. One PoW per transaction therefore means requiring a queue of transactions: If there is no queue and if users wish to execute x many transactions each hour, then x many transactions will be added to the chain each hour, and it will be the rate at which users wish to execute transactions that determines how fast the chain grows.

6. It is often asked whether other forms of permissionless blockchain will have more impact than cryptocurrencies. Once one has the latter providing a tamper-proof ledger, this can be used for other applications. Without using a cryptocurrency, however, the task of motivating users to follow protocol will have to be achieved by means other than payment in currency.

7. *See* https://www.cbeci.org/.

8. For a more detailed analysis, we refer the reader to [DW13].

9. Of course, it might still be a good idea to increase the rate by a lower factor.

10. It is a simplification to talk only in terms of the number of transactions. Transaction complexity is also a factor.

11. While this is not presently realistic, it could soon be feasible through the use of smart contracts such as Truebit [TR18].

12. There are a number of ways to implement these details. The Lightning Network is built specifically for Bitcoin, which means that it is designed with the particular functionalities provided by the Bitcoin scripting language in mind. For the sake of simplicity, however, we shall consider building a payment channel on top of a blockchain with a Turing complete scripting language.

References

[BSBHR18] Eli Ben-Sasson, Iddo Bentov, Yinon Horesh, and Michael Riabzev, *Scalable, transparent, and post-quantum secure computational integrity*, 2018. https://eprint.iacr.org/2018/046.

[DW13] Christian Decker and Roger Wattenhofer, *Information propagation in the Bitcoin network*, Ieee p2p 2013 proceedings, 2013, 1–10.

[ES14] Ittay Eyal and Emin Gün Sirer, *Majority is not enough: Bitcoin mining is vulnerable*, International Conference on Financial Cryptography and Data Security, 2014, 436–454.

[GKL15] Juan Garay, Aggelos Kiayias, and Nikos Leonardos, *The Bitcoin backbone protocol: Analysis and applications*, Advances in Cryptology—EUROCRYPT 2015. Part II, Lecture Notes in Comput. Sci., vol. 9057, Springer, Heidelberg, 2015, 281–310, DOI 10.1007/978-3-662-46803-6_10.MR3344957.

[KBKS18] Uri Klarman, Soumya Basu, Aleksandar Kuzmanovic, and Emin Gün Sirer, *bloXroute: A scalable trustless blockchain distribution network whitepaper*, IEEE Internet of Things Journal (2018).

[KOH19] Daniel Kraus, Thierry Obrist, and Olivier Hari, *Blockchains, smart contracts, decentralised autonomous organisations and the law*, Edward Elgar Publishing, 2019.

[LPR20] Andrew Lewis-Pye and Tim Roughgarden, *Resource pools and the cap theorem*, submitted (2020).

[Lyn96] Nancy A. Lynch, *Distributed algorithms*, The Morgan Kaufmann Series in Data Management Systems, Morgan Kaufmann, San Francisco, CA, 1996. MR1388778.

[N+ 08] Satoshi Nakamoto et al., *Bitcoin: A peer-to-peer electronic cash system*, 2008.

[PSS17] Rafael Pass, Lior Seeman, and Abhi Shelat, *Analysis of the blockchain protocol in asynchronous networks*, Advances in Cryptology—EUROCRYPT 2017. Part II, Lecture Notes in Comput. Sci., vol. 10211, Springer, Cham, 2017, 643–673, DOI 10.1007/978-3-319-56614-6. MR3652143.

[PD16] Joseph Poon and Thaddeus Dryja, *The Bitcoin Lightning Network: Scalable off-chain instant payments*, 2016.

[Ren19] Ling Ren, *Analysis of Nakamoto consensus*, Cryptology ePrint Archive, Report 2019/943 (2019). https://eprint.iacr.org.

[SZ15] Yonatan Sompolinsky and Aviv Zohar, *Secure high-rate transaction processing in Bitcoin*, Financial cryptography and data security, Lecture Notes in Comput. Sci., vol. 8975, Springer, Heidelberg, 2015, 507–527, DOI 10.1007/978-3-662-47854-7_32. MR3395041.

[TR18] Jason Teutsch and Christian Reitwießner, *Truebit: A scalable verification solution for blockchains*, 2018.

[W+ 14] Gavin Wood et al., *Ethereum: A secure decentralised generalised transaction ledger*, Ethereum project yellow paper **151** (2014), no. 2014, 1–32.

Logical Accidents and the Problem of the Inside Corner

MICHAEL C. DUDDY

Introduction

Mathematics as an expression of the human mind reflects the active will, the contemplative reason, and the desire for aesthetic perfection. Its basic elements are logic and intuition, analysis and construction, generality and individuality.

—Richard Courant, *What Is Mathematics?*
An Elementary Approach to Ideas and Methods

As architectural practice has come to fully embrace digital technology, the algorithmic programs that control the design processes have become an important concern in architectural education. Such algorithms enable the architect to input data as variables into a logical ordering system which in turn outputs representations that correspond to the parameters provided. In fact, the presence of algorithmic methods in the form of rules of design can be traced back to Vitruvius, famously illustrated by the ideal proportions of the human figure.[1] His analogy inspired a full set of formulaic rules for the proper dimensions of the Doric temple that when reinterpreted by Alberti fourteen centuries later provided the rules that impacted architectural practice into the twentieth century. If the rules were followed, according to Alberti, then all parts of the building would correspond and one could determine the proportional relations of the entire building from simply taking the dimensions of an individual part, and the building would embody *perfection*. As computational procedures and artificial intelligence displace human reason in the production of architectural form, will its geometry become merely *exceptional*[2] and epistemologically beyond the capability of human reason—that is, a geometry we cannot understand—leading to the question of whether future architecture will be capable of achieving a perfection we can apprehend? The complex building geometries made possible by computational design bring relevance to the question of whether there is a geometry inherent to architecture: a geometry that is not applied from outside the discipline of architecture. If so, what distinguishes an architectural geometry from other types of geometry, notably the geometry of mathematics or the geometry of engineering, and what is the foundational logic of this geometry?

Through a close analysis that focuses primarily on the condition of the inside corner, this paper investigates how the logical consequences of a simple system of linear or gridded repetitions that underlie the foundational logic of the architecture that preceded the arrival of the digital turn—a type of *analog algorithm*[3]—lead to complex and seemingly inconsistent conditions when the system meets at the corner. Such consequences are manifested either as *accidents*—visually unresolved conditions that are nevertheless consistent with and conform to the logic of the system—or as *interventions* where the architect violates the system in order to resolve the condition by means of an aesthetic judgment applied from outside the system. Accidents are considered manifestations of the consistency or "truth" of the system, while interventions are understood as inconsistencies imposed on the system through aesthetic judgments. It follows, therefore, that perfection is not a truth but an aesthetic judgment and that such judgments are inherent to the foundational logic of architectural geometry.

In *The Nature and Meaning of Numbers,* the nineteenth-century mathematician, Richard Dedekind, describes a formal system in mathematics as a rule-based discipline that is abstract and rigorous, and whose rules must be executed in a logical order such that no contradictions are encountered and consistency is ensured. Furthermore, the system must stand without meaning, that is, it must be independent of any references to space and time.[4] Similarly, computational designers Achim Menges and Sean Ahlquist define an algorithm in computational design as "a set of procedures consisting of a finite number of rules, which define a succession of operations for the solution of a given problem. In this way, an algorithm is a finite sequence of explicit, elementary instructions described in an exact, complete yet general manner."[5] Accordingly, the foundational logic of computation, just as it is in mathematics, is a discrete set of ordered rules. Yet despite his assertion,

Dedekind acknowledges that there is a point where human intuition participates, as with the reception of an "elegant" proof of modern geometry that satisfies the mind.[6] Because the practice of geometry in architecture is directed toward the harmony of material and space, satisfaction here results when the building appears resolved such that "nothing can be added or taken away except for the worse," as Alberti famously said.

For this study, we will consider the formal system of grids, repetitions, and alternations—a simple system underlying the fundamental modular blocks from which architecture has been traditionally generated. Such a simple system produces what have been described as *metric patterns* that differentiate this simple system from their random environment.[7] For example, the use of identical columns aligned according to a standard interval can be traced back to the earliest buildings where efficient construction required the economy that accompanies repetition. And with the initial publication of Alberti's *De Pictura* in 1433 (1988), the grid entered the architectural canon as the presiding principle to organize experiential space. As Alberti demonstrated, the grid accommodated the orthogonalization of metric space that was essential for the realization of the geometrical construction of linear perspective. Furthermore, Alberti's gridded "veil" provided a framework to translate points in space onto a surface. Henceforth, space could be accurately represented using a rigorous geometrical method and did not rely on the eye of the artist. Tzonis and Lefaivre describe this orthogonalization as an aspect of *taxis*,[8] which, they write, is

> the ordering framework of architecture, divides the building into parts and fits into the resulting partitions the architectural elements, producing a coherent work. . . . . Taxis contains two sublevels, which we call schemata: the grid and the tripartition. The grid schema divides the building through two sets of lines. In the rectangular grid schema, which is one of the most used in classical architecture, straight lines meet at right angles.[9]

They further note that taxis determines the limits of the metric pattern, that the pattern has to begin as well as end at prescribed points.[10] The prescribed points of commencement and termination of a simple repetitive system, the subject of this paper, is the right angle where the system of periodic repetition changes direction, namely, the corner.

What Is a Logical Accident?

Following an algorithmic procedure can lead to what appear as inconsistencies and a loss of order. As Stephen Wolfram illustrates with his Cellular Automata, the execution of a simple set of rules can lead to complex results.[11] Many of his programs lead to readily recognizable patterns, while others present random patterns seemingly devoid of logic when in fact they are consistent with a prescribed set of rules (Figure 1). In architecture, logical accidents occur when a simple repetitive system meets the corner; these accidents are most evident at the inside corners. In typical early Renaissance courtyards such as the Ospedale degli Innocenti or Santa Croce, both in Florence, identical columns fall at the points of intersection of the lines that determine a perfect grid plan. As a result, the arches at the corners appear to collide and the spandrels appear to cut off arbitrarily, leaving fragmented shields hinged at 90 degrees (Figure 2). This condition is a logical accident that results by conforming the columns to the grid. Greek architects confronted the problem of turning the corner of an Ionic portico. The frontality of the Ionic capital led the architects to merge the volutes of the orthogonal sides to create an outside corner that is diagonally oriented to make the turn. However, the inside corner, appearing as a curious fragment, was nevertheless consistent with the logic of intersecting two Ionic capitals at perpendicular orientations (Figure 3).

In architecture, we can identify three classes of simple systems: (1) single-order planar systems, (2) double-order linear systems, and

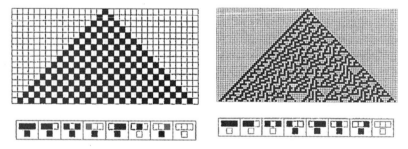

FIGURE 1. Two examples of Stephen Wolfram's cellular automata following different programming rules. Rules such as Rule 250 output repetitious patterns, while rules such as Rule 30 generate irregular complex patterns. [Stephen Wolfram, *A New Kind of Science.*]

FIGURE 2. Two fifteenth-century Florentine interior courtyards, Santa Croce (left) and Ospedale degli Innocenti (right), show that a rigorous module of columns leads to compromised spandrel conditions at the inside corner.

FIGURE 3. Turning the corner with an Ionic column presented architects with a difficult formal challenge. At the outside corner, the volute was adjusted on a 45° angle to make the turn, while on the inside corner, the two volutes merely collided, as can be seen in the Palazzo Barbarano of Palladio. [Drawing by Palladio, Book IV.]

(3) centric systems. Single-order planar systems are uniform grids where the consistency of the corner is maintained, usually the outside corner.[12] At the Beinecke Rare Book Library at Yale in New Haven, Connecticut, Gordon Bunshaft carefully sustained the modular grid system at the outside elevations of the building to express its

FIGURE 4. To maintain the geometric purity of the Vierendeel grid on
the exterior of the Beinecke Library at Yale University, architect Gordon
Bunshaft (Skidmore, Owings, and Merrill) had to accommodate an interior
corner that seemingly violated the rigor of the grid.

Vierendeel frame. In order for the outside corners to comply exactly
with the gridded system, the inside corners are compressed (Figure
4). Clearly in this case, the accident of the inside corner is the result
of maintaining the consistency of the outside; the rigor of the system
expressed on the outside takes precedence over the visual resolution of
the inside corner. Similarly, in Santo Spirito, Brunelleschi takes up the
modular system he introduced in the loggia of the Ospedale degli Innocenti. Here both the nave and the transept are organized according
to the rule of gridded repetition that maintains a clear logic and consistency throughout the interior as repetitive squares. At the corner
on the exterior, however, the semicircular chapels overlap, creating
a diagonal condition, which on the exterior yields an odd collision of
windows at the inside corner that appears accidental and unresolved
(Figure 5). Tzonis and Lefaivre write that "a building made out of a
single homogeneous division [such as a grid] runs no risk of violating
taxis. Metaphorically we might call it a tautology. It is itself; it contains no element that can contradict it."[13] However, while this may be
true of simple planar systems in mathematical geometry, accounting
for the material thickness of building inherent to architectural geometry means that either the interior or exterior corner will appear as
an accident.

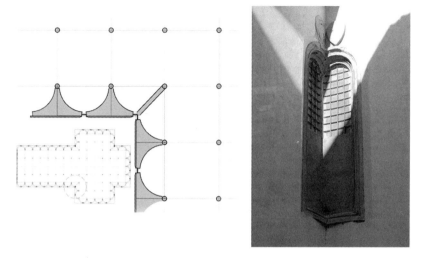

Figure 5. The rigor of Brunelleschi's plan grid at Santo Spirito led to an odd collision of chapel windows on the exterior wall.

Double-Order Linear Systems

A rigorous linear repetitive ordering system, conforming to a primary order, is often accompanied by a corresponding secondary system that follows the same sequential repetition as the primary system, but at a different interval. These systems usually appear as alternation between an object and a space, say a mullion and a window or a column and an interstitial void. For example, when we consider a simple algorithm that involves a sequential repetition of architectural elements, such as the peristyle of a Doric temple, turning the corner becomes a significant formal problem. In the Roman Doric system, the corner triglyphs of the frieze are centered over the column, leaving a fragment of a metope to turn the corner. This is a logical accident that results from the systematic consistency of the primary order of the peristyle, which allows all the columns to be identical and equally spaced. The resolution of the outside corner in the Roman peristyle follows a sequential logic that requires no aesthetic intervention. In the example of Palladio, this accident is especially pronounced in the ornamented metopes of the Palazzo Chiericati, where only minor fragments of the shields turn the corner (Figure 6). The Greek system, on the other hand, mandates

FIGURE 6. For the Roman Doric order, corner triglyphs were centered on the columns below, leaving a small metope fragment at the corner of the frieze. This odd condition is exacerbated in Palladio's ornamental shields on the Palazzo Chiericati.

that the frieze turn the corner with a full triglyph, and in the ideal condition, such as the Parthenon, all triglyphs and metopes should be consistent and equal. This necessitates off-centering the corner column and enlarging its diameter to account for the triglyph's acentric alignment (Figure 7). In this case, the frieze takes precedence as the primary order, and the intercolumniations are aesthetically adjusted to accommodate the order. The Greek corner is the consequence of an aesthetic judgment that intervenes from outside the logic of the system in order to visually resolve the condition.

In the case of both the Greek and the Roman Doric, there is a primary and secondary order, both of which are directly corresponding, but at different intervals. This I will call a double-order linear system.

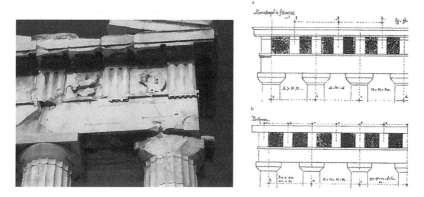

FIGURE 7. The rules of the Greek Doric order required that the corner of the frieze be turned by a full triglyph. This meant that either the regular spacing of the columns was maintained at the sacrifice of the last metope (drawing above), or that the column spacing at the corner be reduced to permit a consistent frieze (drawing below). (Drawing from Durm 1892. This image is from Tzonis and Lefaivre 1986.)

At the Seagram Building, Mies van der Rohe maintains his system of columns arranged on a perfectly square grid as his primary order. Accordingly, the secondary order of curtain wall mullions follows an interval such that one will always fall on the centerline of the column. When it reaches the outside corner, the curtainwall stops mid-column and does not make the turn, an aesthetic judgment that avoids a fragment of curtainwall at the turn as we saw in the Roman peristyle, allowing the column to be expressed the entire height of the building. However, by maintaining the rigor of this column-mullion relationship, the inside corner at the rear of the building succumbs to the logical consequence of shortened modules where the planes intersect (Figure 8). Mies seemed uncomfortable with this accident within his system, intervening with an entirely different approach to the inside corner at One Charles Center in Baltimore.

San Giorgio Maggiore in Venice reveals a seemingly resolved system between the interior column positions and the cornice (Figure 9). For the cornice in the interior of San Giorgio, Palladio successfully combines square inner and outer corners as he abuts full columns and half piers. Here, he has centered the primary order of the columns and piers

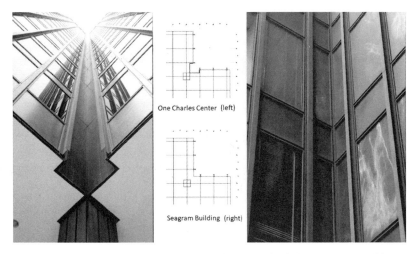

One Charles Center (left)

Seagram Building (right)

FIGURE 8. The rigor of the equilateral column grid of the Seagram Building caused the inside corner condition to appear off the module. Clearly this made Mies uncomfortable, as he attempted to "correct" it at One Charles Center. (Photo of One Charles Center: Rachel Sangree, Johns Hopkins University.)

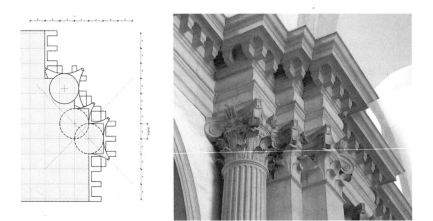

FIGURE 9. The modillions of San Giorgio are spaced such that the module placement leads to perfect squares at both the inside and outside corners. However, the diagram shows that Palladio had to displace the column off his system so that its volute could appear full.

FIGURE 10. In Il Redentore, pilasters comply to the module, while the modillions in the cornice create an odd condition at the inside corner.

on a modillion and has properly proportioned the cornice to allow the secondary order of modillions and voids to appear consistent. However, it can be noticed that in order to maintain the rigor of the metric system of the cornice, Palladio has shifted the position of the column off his modular system so that a complete volute can be expressed at the inside corner. An intervention on the logic of the system leads to the appearance of resolution.

At Il Redentore, also in Venice, Palladio confronts a different formal challenge to his system. At the crossing, Palladio placed the modillions in the cornice such that they form perfectly square recesses between them; these modillions directly relate to the columns and piers below, but this time he centered the primary order of half columns and piers on the void between the modillions (Figure 10). The position of the modillions and the dimension of the cornice ensured that the outside corners always would be square. However, at the inner face of the transept wall, in order to accommodate the smooth blending of a half column and a half pier, the inside corners appear enlarged and a full modillion appears awkwardly in the corner. The analysis shows that the centerlines of the columns are shifted so that a full half column or pier is achieved, and the cornice intervals that consequently accommodate this shift and correspond to the rigor of the system generate an awkward condition at the inside corner: an accident inherent to the system.

Centric Planar Systems

Centric planar systems form the third category of simple systems in architecture. This is a specific type of aesthetic intervention that in general follows an identifiable set of rules to achieve visual resolution. Here adjustments to the condition of the inside corner of alternating elements can be seen in courtyards. These conditions are controlled by the overall enlargement or reduction of the perimeter walls of the space with respect to the center. We remember that at Santa Croce the columns remained faithful to the grid while the spandrels collided awkwardly above (Figure 11a). Architects were sensitive to this condition and to varying degrees shifted the grids by expanding or contracting the space. For the purposes of this analysis, we will consider the absolute grid of the type of *cortile*, an internal court surrounded by an arcade, at Santa Croce as the reference grid from which we will examine offset conditions. The Greek letter ρ (rho) is chosen here to signify the width of the pilaster or column, and the degree of the offset is designated by some multiple of ρ. The position of each grid is denoted by the x and y parameters of $(\rho x, \rho y)$ where the reference grid corresponds to $(\rho 0, \rho 0)$, and the degree of the offset is described by the percentage of the width of the column or pilaster that it is offset.

In the example of the Palazzo del Te near Mantua, Giulio Romano offset the grid and placed the courtyard wall such that the inside corner was met by two full pilasters (Fig. 11c). Here the offset of the centerlines of the grid is given by the parameter (ρ, ρ) where the offset grid has shifted one full pilaster width from the reference grid in both directions. The corner appears resolved; the end bays and their bracketing pilasters appear complete, and the intersecting entablatures appear to be fully supported in both directions. In the Teatro Olimpico in Vicenza, as in many other examples illustrated in Palladio's *Four Books*, the resolution of the corner follows an approach similar to Romano's except here a pier or fragment has been added to hold the corner (Figure 11d). In this case, the parameters approximate $(1\frac{1}{2}\,\rho, 1\frac{1}{2}\,\rho)$, and a space is left between the adjacent pilasters. Again, the corner appears resolved and complete; the increased complexity

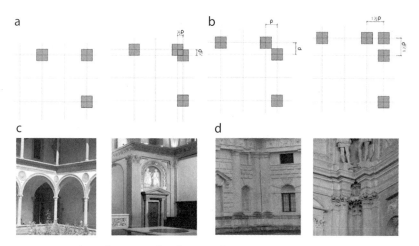

FIGURE 11. The columns in the cloister of Santa Croce [11a] lie on an equilateral orthogonal grid, causing the shield in the corner to be truncated. In the Old Sacristy [11b], Brunelleschi has offset the corner pilasters by $\frac{1}{2}\,\rho(x)$, $\frac{1}{2}\,\rho(y)$, so that the cornice and arches above will appear supported. For the Palazzo del Te [11c], Giulio Romano closes the corner with two full pilasters surmounted by a full triglyph. Palladio goes a step further in the Teatro Olimpico [11d], where a quarter pilaster is flanked by full pilasters (displacement $1\frac{1}{2}\,\rho$, $1\frac{1}{2}\,\rho$).

foretells the Baroque configurations that would follow in the ensuing decades. In the example of the Old Sacristy of the Basilica of San Lorenzo in Florence, Brunelleschi was especially challenged by his pure sphere/cube geometry. Here, the purity of the geometry combined with the dimensions of the stone surrounds of the arches mandated that he corner his perfect square with half pilasters corresponding to a parametric relation of ($\frac{1}{2}\,\rho$, $\frac{1}{2}\,\rho$) (Figure 11b). In each of these cases, a specified symmetrical offset gives a controlled resolution to the corner condition.

Rigor and Intervention

The rigor of the spatial organization of the Pazzi Chapel, near the Basilica di Santa Croce, is well documented,[14] yet even a cursory inspection of what appears as the rigorous placement of pilasters across the interior walls reveals curious conditions where pilaster fragments

Figure 12. Pilaster fragments occur at the typical interior corners of the Pazzi Chapel.

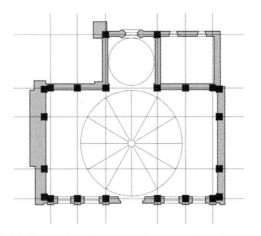

Figure 13. Highlighting the pilasters in the Pazzi Chapel as columns reveals the displacement Brunelleschi used at the corners.

either hold the corners or converge with full pilasters to form the interior corners (Figure 12). When a diagram of full piers is generated to mark the positions of the pilasters, the formal system can be readily discerned (Figure 13). The *scarsella* (a small apse), for example, reveals that a tiny fragment of pilaster is consistent with the logical system that corresponds to "piers" located on the grid of a square plan (Figure 14). Similar to the Old Sacristy, the entablatures appear to be supported

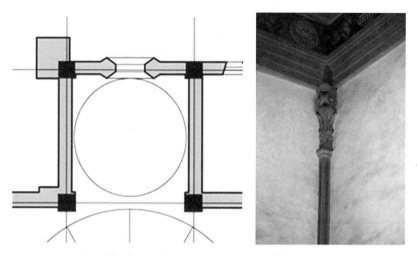

FIGURE 14. The odd pilaster fragments in the scarsella are the result of columns that fall on the grid: a logical accident.

at the corner, and the stone surrounds of the arches above are visually carried to the floor by these pilaster fragments.

In the main room of the chapel, however, there is an apparent intervention on his system: the row of pilasters on the lateral walls are offset from the dominant grid that is generated by the dome by the width of a full pilaster (ρ), presaging a formally challenging condition at the corners. Specifically, the imaginary pier that corresponds to the pilaster fragment is offset asymmetrically from the dominant system by (ρ, 1/6 ρ) (Figure 15). The full pilaster offset in the x-direction could be justified since it allows for a full stone rib to balance a similar full rib that supports the dome. The offset in the y-direction, however, is more problematic. The position of these "piers" offset by 1/6 ρ results in a compromise to system due to the reduction of the module at the end bays of the lateral walls sufficient to account for the insertion of the pilaster fragment. Brunelleschi has again intervened on the logic of his system by introducing these pilaster fragments that do not conform to the system. However, just as in the scarsella, the entablatures at the end lateral walls of the main room appear supported at their ends, and the stone arch that surmounts the side wall beneath the end vaults would not appear to be supported below without them.

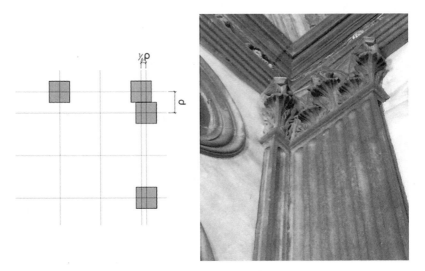

FIGURE 15. The odd fragments in the main room of the Pazzi Chapel are displaced from the grid by $(1/6\ \rho,\ \rho)$ to visually support the entablature and arch above: an aesthetic intervention.

The Logical and the Aesthetic

The problem of the corner is a problem of architectural geometry. In a simple repetitive system, the corner can be expressed as a logical accident which appears as an imperfection that is nonetheless inherent to the system, or it can be expressed as an intervention that "resolves" the accident. Since the resolution of the accident falls outside the formal system, we can conclude that the intervention is an expression of an aesthetic judgment, where the corner, although appearing resolved, is actually a compromise to the rigor of the system: the presence of an aesthetic judgment constitutes an inconsistency to a system for the sake of achieving perfection. Thus, when an aesthetic judgment is asserted, two differing modes of reason confront one another—one of logic and one of intuition; one of intellect and one of feeling; one of truth and one of beauty. While formal rigor is fundamental to the consistency of architectural geometry, so too are the interventions of aesthetic judgments that make the building appear resolved toward perfection. And at the philosophical level, perfection—the unity of reason and judgment—is ultimately an epistemological question as much as it is a formal problem.

The French mathematician Henri Poincaré contemplates the emotive sensibility within the mathematician who apprehends in a mathematical demonstration a quality which to most would seem to only interest the intellect. "But not if we bear in mind the feeling of mathematical beauty," he writes, "of the harmony of numbers and forms of geometrical elegance. It is a real aesthetic feeling that all true mathematicians recognize, and this is truly sensibility."[15] Such a sentiment pervades the world of science as well, where David Orrell observes that the "truth" of science is often obfuscated by the quest for elegance.

> The way that we see the world has been shaped by the traditional scientific aesthetic. We look for reductionist theories that can break the system down into its components. We seek elegant equations that can be rigorously proven using mathematics. We aim for a unified, consistent theory. We celebrate symmetry, clarity, formal beauty. . . . What we mean by "good science" is strongly related to what we (i.e., scientists) mean by beauty.[16]

In the experiential geometry of architecture, logic and beauty often stand in opposition to one another. The resolution of the corner invokes a type of geometrical reasoning whose quest is for aesthetic resolve as much as it is for formal truth. This quest is a uniquely human enterprise intrinsic to our being in the world.

Architecture is inherently a geometrical construction; its geometry is the physically constructed expression of our being in the world. It is here that we find a place where the geometry of mathematics (desire for truth) coincides with the geometry of architecture (desire for beauty) identified by Richard Courant in the epigraph of this paper as the "desire for aesthetic perfection." Prior to digital technologies, the rules and principles of architectural geometry were formulated by the human mind: it provided the logical system that was used to construct our habitable space as we intuit it. Its laws and principles were constituted by the confluence of logical reasoning and aesthetic intervention. With the arrival of digital technologies, however, we are moving away from intuition and relegating the production of form and space to the algorithm; as we do so, we surrender the rules of designing our visual environment to the logic of the machine, a device bereft of intuition and aesthetic judgment. The computational shift that has promoted the design process over the formal object and moved the architect out of a central role in the design process has generated architecture that surpasses the intellect of the architect.[17] Hence, will our future interest in architectural form and space lie in its exceptionality instead of in its perfection?

Notes

1. Hersey (1976: 192). Hersey elaborates on the notion of a Vitruvian algorithm, a formula using words and numbers to signify a building, specifically a temple: "... Vitruvius was able to 'reduce' a temple to an algorithm. Such algorithms compressed, stored, and transmitted the temples they described, making them reproducible across separations of time and space. One also infers that Vitruvius would have defined these algorithms as 'signifieds,' while the temples themselves were 'signifiers.' Hence a given temple's meaning is the algorithm describing it." Unlike the computational algorithm, however, the basis of Vitruvius' algorithm, the "cubic procreator," was steeped in Pythagorean meaning and neo-Platonic mysticism.

2. Jones (2014: 39). "Perfection and exception differ in their metrics. Perfection requires a well-established formal system in order to be teleologically recognizable, and it prizes the skill for navigating that territory. Exceptionality requires (only) enough understanding to identify the difference that invention cherishes, and rewards the open attitude that leads to discovery."

3. The concept of *accidents* used here is based on Aristotle's assertion that appears first in the *Posterior Analytics* and later in the *Metaphysics* that an attribute inherent to an object, or in our case, a system, "attaches to a thing in virtue of itself, but not in its essence." (Aristotle 2001: 1025a, 30–32). In Aristotle's example, the fact that the angles of a triangle are equal to two right angles is inherent to a (Euclidean) triangle, but not the essence of triangle. Hence, the attribute is true, but also an accident.

4. Richard Dedekind writes, "In speaking of arithmetic as a part of logic I mean to consider the number-concept entirely independent of the notion of intuitions of space and time, that I consider it an immediate result from the laws of thought ... numbers are free creations of the mind." (Dedekind 1963: 31).

5. Menges and Ahlquist (2011: 11).

6. Dedekind (1963: 38). "All the more beautiful it appears to me that without any measurable quantities and simply by the finite system of simple thought-steps man can advance to the creation of the pure continuous number-domain; and only by this means in my view is it possible for him to render the notion of continuous space clear and definite."

7. Tzonis and Lefaivre (1986: 118) "Repetition, or periodic alternation of compositional units, makes the work stand out in relation to the amorphous random spaces that characterize the surrounding world. Rhythm employs stress, contrast, reiteration, and grouping in architectural elements. By using these aspects of formal organization, *metric patterns* emerge."

8. Tzonis and Lefaivre describe *taxis* as the overriding formal pattern of the work.

9. Tzonis and Lefaivre (1986: 9).

10. Tzonis and Lefaivre (1986: 129).

11. Wolfram (2002).

12. The CBS Building in New York offers an example where the inside corner maintains the rigor of the grid. In this building, Eero Saarinen doubled the mass of the outside corner column to ensure a perfect gridded module on the inside.

13. Tzonis and Lefaivre (1992: 14).

14. See Battisti (2012: 222–229).

15. Poincaré (2004: 37).

16. Orrell (2012: 237).

17. Terzidis (2011: 94). "What makes algorithmic logic so problematic for architects is that they have maintained the ethos of artistic sensibility and intuitive playfulness in their practice. In contrast, because of its mechanistic nature, an algorithm is perceived as a non-human creation and therefore is considered distant and remote."

References

Ackerman, James S. 1991. *Distant Points: Essays in Theory and Renaissance Art and Architecture.* Cambridge, MA: MIT Press.

Alberti, Leon Battista 1988. *On the Art of Building in Ten Books.* Trans. Joseph Rykwert, Neil Leach, Robert Tavernor. Cambridge, MA: MIT Press.

Aristotle 2001. *The Basic Works of Aristotle.* Richard McKeon, Ed., New York: Modern Library.

Battisti, Eugenio 2012. *Filippo Brunelleschi.* New York: Phaidon Press.

Carpo, Mario 2017. *The Second Digital Turn: Design beyond Intelligence.* Cambridge, MA: MIT Press.

Choay, Francoise 1997. *The Rule and the Model: On the Theory of Architecture and Urbanism.* Cambridge, MA: MIT Press.

Courant, Richard, and Herbert Robbins 1996. *What Is Mathematics? An Elementary Approach to Ideas and Methods* (1941). New York: Oxford University Press.

Dedekind, Richard 1963. "The Nature and Meaning of Numbers," in *Essays on the Theory of Numbers* [1901]. New York: Dover Editions.

Durm, J. 1892. *Die Baukunst der Griechen.*

Furniari, Michele 1995. *Formal Design in Renaissance Architecture from Brunelleschi to Palladio.* New York: Rizzoli International Publications.

Hersey, George L. 1976. *Pythagorean Palaces: Magic and Architecture in the Italian Renaissance.* Ithaca, NY: Cornell University Press.

Jones, Wes 2014. "Can Tectonics Grasp Smoothness?" in *Log30: Observations on Architecture and the Contemporary City.* Cynthia Davidson, Ed., New York: Anyone Corporation, Winter 2014. pp. 29–42.

Menges, Achim, and Sean Ahlquist (Eds.) 2011. *Computational Design Thinking.* West Sussex, UK: John Wiley & Sons, Ltd.

Mitchell, William J. 1990. *The Logic of Architecture: Design, Computation, and Cognition.* Cambridge, MA: MIT Press.

Orrell, David 2012. *Truth or Beauty: Science and the Quest for Order.* New Haven, CT: Yale University Press.

Palladio, Andrea 1965. *The Four Books of Architecture* [Book IV]. Mineola, NY: Dover Publications, Inc.

Poincaré, Henri 2004. *Science and Method* [1908]. Trans. Francis Maitland. New York: Barnes and Nobel Books.

Serlio, Sebastiano 1982. *The Five Books of Architecture.* New York: Dover Publications, Inc.

Terzidis, Kostas 2011. "Algorithmic Form" in *Computational Design Thinking*. Achim Menges and Sean Ahlquist (Eds.). West Sussex, UK: John Wiley & Sons, Ltd.

Terzidis, Kostas 2016. *Algorithmic Architecture*. Oxford, U.K.: Architectural Press.

Tzonis, Alexander, and Liane Lefaivre 1986. *Classical Architecture: Poetics of Order*. Cambridge, MA: MIT Press.

Wolfram, Stephen 2002. *A New Kind of Science*. Champaign, IL: Wolfram Media, Inc.

Cosmatesque Design
and Complex Analysis

STEVE POMERANTZ

The Cosmati, families of marble workers, decorated altars, thrones, and pavements in many churches during the twelfth and thirteenth centuries in and around Rome, Italy. One aspect of their design, now known as Cosmatesque, consists of large sections of floor covered with mosaics of regular and semiregular tilings. The images in Figure 1 show a regular tiling of squares, a semiregular tiling with triangles and hexagons, and a semiregular tiling with octagons and squares.

Here, we focus on the non-regular tiling—each vertex is not identical—in Figure 2, consisting of hexagons, triangles, and rhombi.

Figure 2 also dem onstrates a central motive in Cosmatesque tiling known as the *guilloche*, a serpentine arrangement of two interlacing bands, thickened to contain a pattern within. Figure 3 shows another example.

The configurations of shapes in a rectangular tiling are often used to tile these circular bands, as illustrated in Figure 4. Moreover, the

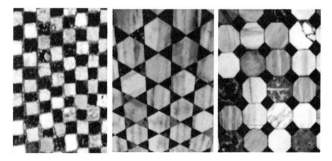

FIGURE 1. Three Cosmatesque tilings. See also color insert.

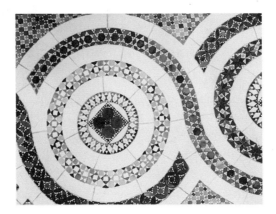

FIGURE 2. A watercolor painting by the author of a floor design; the outer ring consists of a nonregular periodic floor tiling. See also color insert.

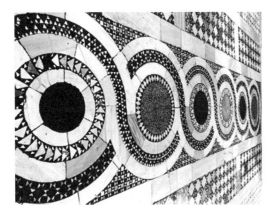

FIGURE 3. Guilloche is a serpentine arrangement of interlacing bands, each containing a pattern. See also color insert.

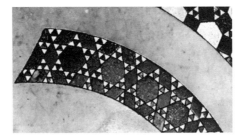

FIGURE 4. Filling in a band with a tiling. See also color insert.

FIGURE 5. A tiling with linear and circular sections.

same design sometimes joins linear sections with circular bands, as in Figure 5.

We will show how to construct a banded tiling through a technique referred to as *conformal mapping* in the theory of complex analysis. The idea is to construct a rectangular tiling and transfer it to the banded region via a function.

The Complex Exponential Function

To illustrate the method, we first review some basic facts about the set of complex numbers. A complex number, often written $z = x + yi$, can be identified with the ordered pair of real numbers (x, y) in the plane, which we call the Argand plane. Here, x is called the real component of z, and y is the imaginary component. The term i denotes the imaginary unit $i = \sqrt{-1}$ so that $i^2 = -1$. The complex number $2 + 3i$ is illustrated in Figure 6.

We can perform arithmetic with complex numbers. Moreover, we can define complex-valued functions—those that act on a complex number to produce another complex number.

In the mid-1700s, Leonhard Euler showed, by considering power series, that

$$e^{yi} = \cos(y) + \sin(y)i$$

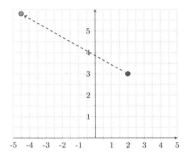

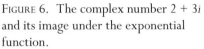

FIGURE 6. The complex number $2 + 3i$ and its image under the exponential function.

This relationship, often referred to as Euler's formula, allows us to define the complex exponential function, e^z: For $z = x + yi$ we get $e^z = e^x \times e^{yi}$, where e^x is the usual real-valued function. In particular, if z is the complex number represented by the point (x, y) in the Argand plane, then e^z is the complex number represented by the point $(e^x \times \cos(y), e^x \times \sin(y))$. Figure 6 shows that the point $2 + 3i$ maps to approximately $-4.5 + 5.8i$.

The complex exponential has applications in many areas of engineering and physics, but here, we use it to create a Cosmatesque tiling. To do this, we must investigate the images of lines under the complex exponential function.

A horizontal line consists of points of the form (x, y_0), where y_0 is constant. These points map to points with coordinates $(e^x \times \cos(y_0), e^x \times \sin(y_0))$. Thus, a horizontal line maps to a radial line; Figure 7 shows two horizontal lines (dotted) and their images under the complex exponential (solid). As x ranges over the real line, the corresponding point moves from near the origin (for large negative values of x) off to infinity (for large positive values of x). When x equals zero, the point lies on the unit circle.

Similarly, a vertical line consists of points of the form (x_0, y), where x_0 is constant. Points of this form map to points with coordinates $(e^{x_0} \times \cos(y), e^{x_0} \times \sin(y))$; this vertical line maps to a circle of radius e^{x_0}, as in Figure 8.

We have thus seen that the complex exponential has the interesting characteristic of mapping the standard horizontal/vertical grid to a grid of circles and their radii; both grids consist of perpendicular sets of lines. In fact, the angle between any two lines in the Argand plane is the same as the angle between their images under the exponential

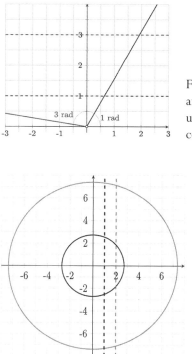

FIGURE 7. Two horizontal lines ($y = 1$ and $y = 3$) along with their images under the exponential map. See also color insert.

FIGURE 8. The vertical lines $x = 1$ (black) and $x = 2$ (red) map to circles with radii e and e^2, respectively, under the complex exponential. See also color insert.

map. Functions that preserve angles are called *conformal maps*; hence, the complex exponential function is an example of a conformal mapping. Note, however, that this mapping does not preserve lengths.

Cosmatesque Patterns from Conformal Maps

Begin with any rectangular tiling pattern lying in a vertical strip. Then apply the exponential map to the vertices of the pattern. The images of these points are the vertices of a tiling pattern within two circular bands (the images of the borders of the vertical strip). Alternatively, several points along each edge can be identified and mapped for a potentially more accurate fit. While diagonal lines are mapped to spirals under the exponential map, the enhanced appearance would be offset by the additional labor of constructing curved pieces.

The left image in Figure 9 shows a non-regular tiling of one hexagon, six triangles, and five rhombi between the vertical lines $x = 1.97$ and

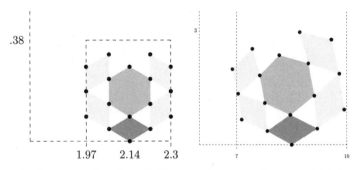

FIGURE 9. A rectangular tiling (left) mapped to a circular tiling (right) under the complex exponential. Note the scales are quite different. See also color insert.

$x = 2.30$. These vertical lines map to circles of radii $e^{1.97} \approx 7.2$ and $e^{2.30} \approx 10$, respectively. The vertices in the tiling map to the corresponding points in the circular band pictured in Figure 9 on the right.

Note that the minimal vertex $(2.14, 0)$ maps to $(8.5, 0)$ while the maximal vertex of the hexagon $(2.14, 0.285)$ maps to $(8.16, 2.39)$. Both of the new points lie on the circle of radius 8.5 approximately 16.4 degrees apart. This angle of rotation corresponds to the difference of y values of the original two points. As stated, the mapping preserves the angles of intersections but skews some of the lengths of individual segments. (If *all* the points were mapped under the exponential function, then the angles would be perfectly preserved. This requires the new edges to be spiral shaped rather than linear.) When the width of the band is small relative to the radii, the difference is not that apparent.

While it is possible to construct this pattern inside a band preserving lengths and angles, there will be visible differences when comparing the behavior along the two edges of the circular band: The longer, outer circle has more space to tile, which forces some asymmetry in the pattern.

When we use $22 \approx 360/16.4$ copies of our initial tile design from Figure 9, we obtain the tiling of a full 360-degree band from Figure 10. The sine and cosine functions have a period of 2π, so we must only draw the original pattern within a strip that is 2π tall. Points outside that range would map to points already identified.

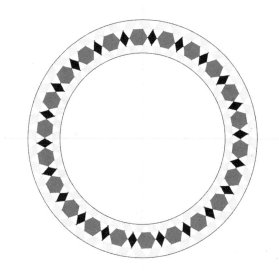

FIGURE 10. The circular tiling resulting from repeated use of the tile in Figure 9. See also color insert.

Starting with a tiling constructed within the rectangle $[a, b] \times [0, 2\pi]$, the exponential function maps the vertices to points inside the circular band with inner radius e^a and outer radius e^b. We then connect the new vertices with straight edges to complete the pattern as illustrated in Figure 10. An applet creating this tiling in GeoGebra can be found at www.geogebra.orglmlgb5fwjzd.

Nullstellenfont

BEN LOGSDON, ANYA MICHAELSEN,
AND RALPH MORRISON

Euclid, in his *Elements*, defined geometric shapes elegantly: a line is length without breadth, a circle consists of all points equidistant from a common point, and so on. Unfortunately, these poetic descriptions are not often the most useful to work with in practice; we instead define shapes using *polynomial equations*, as pioneered by René Descartes. Given a polynomial equation in two variables of the form $p(x, y) = 0$, we can investigate the set of points in the Cartesian plane that make the equation true. For instance, the equation $y - 2x - 1 = 0$ describes the line with slope 2 and y-intercept 1 pictured on the left in Figure 1, and the equation $x^2 + y^2 - 1 = 0$ yields the unit circle shown on the right in Figure 1.

The correspondence between geometric shapes and algebraic equations inspires many questions. What can we glean about the shape from the equation? How do algebraic operations correspond to geometric operations? For instance, if we multiply the defining polynomials of the line and circle given earlier, we create the new polynomial equation

$$(y - 2x - 1)(x^2 + y^2 - 1) = 0$$

A point in the plane satisfies this equation if and only if at least one of the factors of the left side is 0, so the shape defined is the set of all points lying on the line or the circle or *both*, i.e., this equation describes the *union* of the line and the circle, as pictured in Figure 2. Thus, an algebraic operation (multiplication) on equations corresponds to a geometric operation (taking a union) in the plane.

Algebraic geometry is the study of shapes defined by polynomial equations (and vast generalizations thereof!). Such a shape is called a *variety* (or sometimes an *algebraic set*). As with lines and circles already

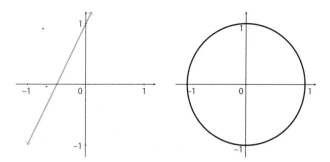

FIGURE 1. A line and a circle described by polynomial equations.

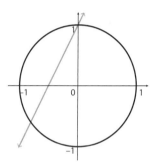

FIGURE 2. The union of the line and the circle obtained by multiplying their respective equations.

encountered, polynomial equations in two variables correspond to shapes in the plane. Polynomial equations in three variables give rise to shapes in three dimensions, and in general, polynomial equations in n variables define shapes in n-dimensional space.

Many cool shapes in the plane, such as the heart in Figure 3, can be described using polynomial equations and are thus varieties. Granted, the more complex a shape is, the bigger the polynomial might be; for instance, the heart corresponds to the equation

$$x^8 + 240x^6y + 256x^4y^3 + 8{,}116x^6 + 70{,}464x^4y^2$$
$$+ 67{,}584x^2y^4 + 16{,}384y^6 + 597{,}312x^4y$$
$$+ 1{,}818{,}624x^2y^3 + 589{,}824y^5 + 15{,}918{,}304x^4$$
$$+ 6{,}081{,}792x^2y^2 + 2{,}899{,}968y^4 - 471{,}039{,}744x^2y$$
$$- 71{,}958{,}528y^3 - 3{,}937{,}380{,}544x^2 - 246{,}497{,}280y^2$$
$$+ 4{,}261{,}478{,}400y - 10{,}061{,}824{,}000 = 0$$

The third author of this article taught a course on algebraic geometry, during which the first two authors conceived of a final project to

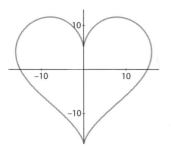

FIGURE 3. A heart defined by a polynomial equation.

build varieties that look like any given string of letters. For instance, one could ask for a polynomial $f(x, y)$ such that the set of points satisfying $f(x, y) = 0$ looks like "Hello World" (spoiler alert: Figure 9 contains such a variety). The use of polynomial equations to create fonts is not new: certain polynomial-defined shapes, called Bézier curves, are used to render most fonts, where each letter is made up of many pieces of curves glued together. Our project was a bit more ambitious: We wanted each entire phrase to be described by a single polynomial equation.

We will tell you how we pulled this off, from building individual letters of the alphabet to stringing the letters together to create any phrase! We named our typesetting tool Nullstellenfont, inspired by David Hilbert's famous Nullstellensatz (or "zero-locus theorem"), which characterizes when different polynomials describe the same variety.

Building Blocks

Most letters in the English language require line segments. The closest variety to a line segment is an entire line, given by an equation of the form $ax + by + c = 0$. Unfortunately, lines stretch on forever, so they are not quite what we wanted for creating nice, bounded letters. We realized that a thin ellipse appears like a line and has the added benefit of having thickness. We drew ellipses in the plane using GeoGebra, which produced the necessary polynomial equations.

To ensure that the letters were roughly the same size, we placed the letters inside of a unit square centered at the origin. Finally, we obtained the polynomial for the entire letter by taking the union of the varieties of the pieces of the letter; as we have seen, the union is obtained by multiplying the corresponding polynomials. Figure 4 shows

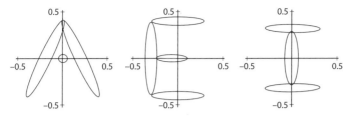

FIGURE 4. The vowels A, E, and I as unions of ellipses.

the vowels A, E, and I as the union of ellipses; for instance, the letter I is defined by an equation $f(x,y)g(x,y)h(x,y) = 0$, where $f(x,y) = 0$, $g(x,y) = 0$, and $h(x,y) = 0$ are the equations yielding the three ellipses.

Beyond Ellipses

Curvier letters provided us with our next challenge. We categorized curvier letters into three groups by the type of curve: small enclosed curves (e.g., B, P, R), big or open curves (e.g., C, D, G, U), and the letter S. For each group, we found curves matching the required shapes.

Small Enclosed Curves

A circle is a simple closed curve, but we wanted something with a little more character. Some curves are much easier to plot using *polar coordinates* r and θ, where r measures the distance to the origin and θ measures the counterclockwise angle from the positive x-axis. For instance, the polar equation $r = 1$ defines the unit circle.

We began with the polar curve given by $r = 1 + 2\cos(\theta)$, pictured in Figure 5. We converted this equation to xy-coordinates by first multiplying by r, resulting in the equation $r^2 = r + 2r\cos(\theta)$. Then, we used the standard transformations $x = r\cos(\theta)$, $y = r\sin(\theta)$, and $x^2 + y^2 = r^2$, to produce $x^2 + y^2 = \sqrt{x^2 + y^2} + 2x$.

Unfortunately, this isn't a polynomial equation, due to the square root. Luckily, varieties only correspond to zeroes of a polynomial, so by rearranging and squaring both sides of our equation, we obtain another equation with the exact same solutions as our original:

$$(x^2 + y^2 - 2x)^2 - x^2 - y^2 = 0$$

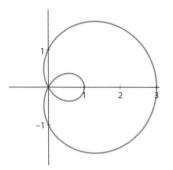

FIGURE 5. Polar curve $r = 1 + 2\cos(\theta)$.

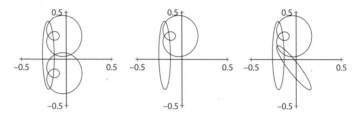

FIGURE 6. The letters B, P, and R consisting of shifted small enclosed curves and ellipses.

We translated and scaled the resulting curve for use as the loop in the letters B, P, and R (Figure 6).

Open Curves

For letters with open curves, we sought inspiration in Wikipedia's "Gallery of Curves," where we encountered the ampersand curve, pictured in Figure 7, defined by

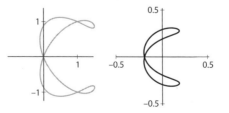

FIGURE 7. The ampersand curve and the letter C.

$$(y^2 - x^2)(x - 1)(2x - 3) - 4(x^2 + y^2 - 2x)^2 = 0$$

We tweaked the coefficients to reshape it until it matched the letter C, shown on the right in Figure 7. Rotating and scaling gave us the letters G and U as well. We also use this curve for D to give a gentler shape than the small enclosed curve.

The Letter S

To create the only remaining letter, S, we initially considered the equation $x^3 - xy^2 - y^3 + y = 0$, pictured on the left in Figure 8. The result was too big to fit in the required unit square, so we scaled this curve to fit by replacing x with sx and y with sy for some scaling factor $s > 1$. By choosing two curves with slightly different scaling factors, we obtained some thickness to the resulting letter, as shown on the right in Figure 8.

Stringing Letters Together

The final challenge was to determine how to combine the letters to form any phrase. As we have seen, we can take the union of two varieties by multiplying their equations, so the only remaining step is to translate each letter in the text horizontally by an appropriate amount.

Of course, we can translate by substituting $(x - n)$ for x in the nth letter in the sequence, starting at 0: the first letter uses x, the second letter uses $x - 1$, etc. This composition ensures that the nth letter is plotted in a unit square centered at $(n,0)$. After shifting each letter by the appropriate amount and multiplying their polynomials together,

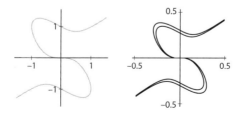

FIGURE 8. The letter S.

FIGURE 9. "Hello World" using the Nullstellenfont.

$$(2.64(x - 1)^2 + 0.0757y^2 + 1.58x - 1.355)$$
$$(0.1(x - 1)^2 + 0.025(x - 1)y + 1.858y^2 + 0.0134x - 0.0027y - 0.0159)$$
$$(0.0355(x - 1)^2 + 1.475y^2 + 1.18y + 0.233)$$
$$(0.0355(x - 1)^2 + 1.475y^2 - 1.18y + 0.233)$$
$$(2.23(x - 2)^2 + 0.274y^2 + 0.89x - 0.03y - 1.73)$$
$$(0.156(x - 2)^2 + 1.156y^2 - 0.0156x + 0.69y + 0.1212)$$
$$(2.23(x - 3)^2 + 0.274y^2 + 0.89x - 0.03y - 2.62)$$
$$(0.156(x - 3)^2 + 1.156y^2 - 0.0156x + 0.69y + 0.1368)$$
$$((x - 4)^2 + y^2 - 0.1225)$$
$$((x - 4)^2 + y^2 - 0.16)(2.73x^2 + 0.16y^2 + 1.632x + 0.2176)$$
$$(2.73x^2 + 0.16y^2 - 1.632x + 0.2176)\,(0.196x^2 + 1.476y^2 - 0.009) = 0$$

FIGURE 10. The polynomial equation for the word "Hello" from Figure 9.

we obtain a single polynomial equation describing the entire text! The variety for "Hello World" appears in Figure 9; the equation for just "Hello" is shown in Figure 10.

The large size of the "Hello" polynomial arises because we multiply the polynomials for the five letters in the word, each of which may have up to four components. Note that D utilizes the open curve and not the small enclosed curve like R. Can you identify the pieces of the polynomial that correspond to the circles making up the letter O?

Want to Make Your Own Text?

If you would like to find a polynomial equation defining a variety that looks like *your* name, or any other string of letters you have in mind, you can use the website https://sites.google.com/williams.edu /nullstellenfont/ to produce the polynomial and/or an image. The code to create that site was written in Sage and can be found on the website link to GitHub.

Further Reading and Acknowledgments

A great introduction to algebraic geometry is *Ideals, Varieties, and Algorithms: An Introduction to Computational Algebraic Geometry and Commutative Algebra* by David A. Cox, Donal O'Shea, and John Little. (This book was used in the course that gave rise to our project.) Also, at a conference in honor of David Cox's retirement, Nathan Pflueger came up with the name "Nullstellenfont" for our project. Thanks, Nathan!

Hyperbolic Flowers

MARIA TRNKOVA

1. Introduction

Bill Thurston introduced several ways to visualize the hyperbolic plane and see some of its properties. He proposed several ways of making it with paper triangles, paper annuli, and fabric pentagons (Thurston 1997). Later he continued this work with Kelly Delp, and they made more complicated surfaces (Delp and Thurston 2011). This approach is described in detail by Kathryn Mann in her notes "DIY Hyperbolic Geometry," which contain many exercises on the topic (Mann n.d.). Other very aesthetic models of hyperbolic surfaces were introduced by Daina Taimina. She suggested crocheting hyperbolic discs (Henderson and Taimina 2001, Taimina 2018). Her suggestion attracted a lot of attention; it was followed by contributions of many other artists, and it even appeared in a TED talk (Taimina n.d.). It is well described in an issue of *Craft: Transforming Traditional Crafts* magazine (Sinclair 2006).

Our approach is inspired by these models, but we take one step further and instead of crocheting solid discs, we make them triangulated. The main difference is that these new models allow us to explain some intrinsic properties of hyperbolic geometry. We also introduce a crochet model of a hyperbolic cylinder. In this note, we want to use the advantages of triangulated models to demonstrate some well-known results of hyperbolic geometry: (i) the relation between edges and angles in a hyperbolic triangle, (ii) tiling of a hyperbolic plane, (iii) circumference growth of a disc, (iv) Nash–Kuiper embedding theorem, and (v) visualization of a cylinder in hyperbolic space. The crochet scheme is given in Figure 9. The models can be used in classes of different levels, from high school to graduate courses, to demonstrate special features of negatively curved spaces as well as to compare them with Euclidean spaces and surfaces.

2. Constructions with Equilateral Triangles

Let us take a collection of equilateral triangles of the same size and connect them along their edges without shifting or stretching. We classify different cases based on the number of triangles meeting at each vertex. We can visualize the process by taking the actual triangles made out of wood, paper, or plastic and trying to glue them together. The simplest case is two triangles, and if we glue them along each edge, we immediately get a *triangle pillow*. Following the same strategy for three triangles meeting at the same point, we get a *tetrahedron*, for four triangles an *octahedron*, and for five triangles an *icosahedron* (Figure 1).

For the next case of six triangles, we realize that all triangles that meet in one vertex must lie in a plane and form a hexagon. If we continue adding six triangles to the previously obtained vertices, we tile the whole Euclidean plane (Figure 2).

We saw that if the number of triangles was less than six, we needed a finite number of triangles and the resulting object was a closed object of *positive curvature*. In the case of six triangles, we got a full plane and the number of triangles was infinite. It has *zero curvature*. If we go even further and demand having seven triangles at each vertex, something very interesting happens. We can start with a single vertex, add seven triangles and get a polygon that cannot be flattened. Nevertheless, we continue to add more triangles to other vertices, and keep this process going. We quickly realize that we again need an infinite number of triangles to satisfy the condition of having seven triangles at each vertex. Unlike in the six-triangle case, the resulting object does not look like a flat plane—it is a *hyperbolic plane* $\mathbb{H}^2$, and it has a *negative curvature*.

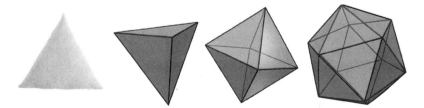

FIGURE 1. Gluing equilateral triangles with a pattern of two, three, four, and five at each vertex. Pictures courtesy of Wikipedia. See also color insert.

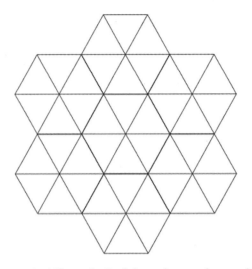

FIGURE 2. Tiling of a Euclidean plane with triangles.

FIGURE 3. The crochet model on the left is a solid hyperbolic disc. The two
models on the right are triangulated hyperbolic surfaces, approximating
hyperbolic discs. They have contours made in a contrasting color to show
their boundaries and better hold shapes. See also color insert.

For a better visualization, we can choose some material and glue the
triangles together. This would be harder to do with a firm material
like paper, wood, or plastic, but this can be nicely done with yarn. We
apply the idea above for creating pieces of hyperbolic plane to crochet
these models. The triangulation of the models allows us to observe
some geometric properties. The crochet models in Figures 3 and 10 are
examples of hyperbolic discs.

3. Topology and Geometry

3.1. TILINGS OF A HYPERBOLIC PLANE

It is very well known that only a few regular polygons can tile the Euclidean plane: triangles, squares, and hexagons. Note that the size of each of the shapes is not fixed. The situation is very different in the hyperbolic case. There are *infinitely* many ways to tile a hyperbolic plane $\mathbb{H}^2$ with *regular* polygons. The most natural way is to tile it with equilateral triangles. Unlike in the Euclidean case, there is a freedom in the number of equilateral triangles at each vertex. Once we choose this number, it determines the size of an equilateral triangle, and it means that the tiling cannot be scaled. If there are k triangles around each vertex of a hyperbolic plane, then every angle of an equilateral triangle must be $2\pi/k$. The smallest number k of equilateral triangles around one vertex that gives a hyperbolic structure is seven. The interesting fact is that the sum of angles in each triangle is $6\pi/k$, which depends on k, $k \geq 7$, and it is always less than π. For Euclidean triangles, the sum of internal angles is always π, and this difference is the main intrinsic property of a hyperbolic plane. Our two models (the right two objects in Figure 3) have seven equilateral triangles around each vertex. They approximate a disc region in $\mathbb{H}^2$ and also demonstrate that this pattern could be uniquely extended to a tiling of the plane. The crochet model looks the same around every interior vertex or triangle.

The hyperbolic plane $\mathbb{H}^2$ can be interpreted by different geometric models. We consider the projective disc model $\mathbb{D}^2$, also known as the Beltrami–Klein model (Stillwell 2005, Thurston 1997). This is an open unit disc with the following metric $d_{\mathbb{D}^2}\,(P, Q)$ in $\mathbb{D}^2$:

$$\cosh d_{\mathbb{D}^2}(P, Q) = \frac{1 - P \cdot Q}{\sqrt{1 - \| P \|^2}\sqrt{1 - \| Q \|^2}},$$

where P, Q are vectors in $\mathbb{D}^2$, $||P||$ is a Euclidean norm of P and is equal to the length of the vector P. Geodesics in this model look like chords of the disc $\mathbb{D}^2$.

We consider a tiling of $\mathbb{D}^2$ with seven equilateral triangles meeting at each vertex. Every angle of such triangle is $2\pi/7 \approx 51.4°$, although it does not look so in the projective disc model in Figure 4. We use the above distance formula to compute coordinates and edge length of the

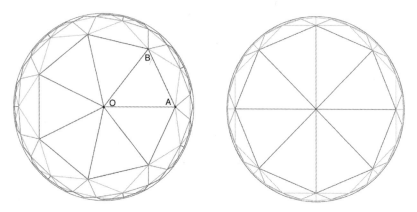

FIGURE 4. Several equilateral hyperbolic triangles in the projective disc model tile the hyperbolic plane with seven and eight triangles, respectively, at each vertex.

equilateral triangle. Let one vertex be at the origin $O(0, 0)$ and the other one A on the x-axis. Then A and the third vertex B should have coordinates $(a_1, 0)$ and (b_1, b_2), which satisfies the equation

$$\cosh d_{\mathbb{D}^2}(O, A) = \cosh d_{\mathbb{D}^2}(O, B) = \cosh d_{\mathbb{D}^2}(A, B)$$

where points O, A, and B can be viewed as vectors. Rewrite these equations once again

$$\frac{1 - O \cdot A}{\sqrt{1 - \|O\|^2}\sqrt{1 - \|A\|^2}} = \frac{1 - O \cdot B}{\sqrt{1 - \|O\|^2}\sqrt{1 - \|B\|^2}} = \frac{1 - A \cdot B}{\sqrt{1 - \|A\|^2}\sqrt{1 - \|B\|^2}}.$$

The equations give four symmetric solutions. One of them is $A(0.797, 0)$, $B(0.496, 0.623)$. We compute the hyperbolic distance between all three points using the same formula, and it is ~ 1.0905, which is the edge distance of a hyperbolic triangle that tiles the hyperbolic plane with seven triangles at each vertex.

In a similar way, one can compute the edge length of an equilateral triangle that tiles $\mathbb{H}^2$ with eight triangles at each vertex. Its edge is ~ 1.5285, and each angle is $\pi/4 = 45°$.

The above calculations demonstrate the fact that there are no similar triangles in the hyperbolic plane: *The angles of a hyperbolic triangle determine its sides, and vice versa.*

Figure 4 shows only a few layers of triangles tiling $\mathbb{H}^2$ around the origin. It appears that the triangles approach the boundary of the unit disc very fast, but this process is infinite, and there are infinitely many triangles in the tiling of a hyperbolic plane. There are four such layers in the crochet models in Figure 3.

3.2. Circumference of a Disc

The ratio of a circumference of a disc to its diameter is constant and is equal to π; remember that $C_{\mathbb{E}^2}(r) = 2\pi r$ in the Euclidean plane. In the hyperbolic plane, the formula for the circumference of a disc with the same radius looks similar: $C_{\mathbb{H}^2}(r) = 2\pi \sinh r$ (Fenchel 1989). The comparison of how fast $C_{\mathbb{H}^2}(r)$ grows with respect to its diameter shows exponential growth (*see* Figure 5). This fact is also visible on the hyperbolic crochet models, where we see that the white and green boundaries on our models are curling in space and cannot be flattened.

3.3. C^1 Embedding

Our models have finite diameter, and it is obvious that they try to occupy some finite size ball in $\mathbb{R}^3$. They look more like curly balls as the radius grows. We want to understand why they look so symmetrically curly.

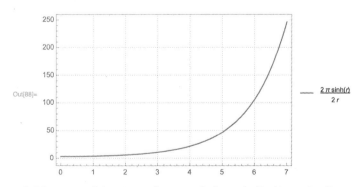

FIGURE 5. The ratio of the circumference of a hyperbolic disc to its diameter as a function of the radius.

The Nash–Kuiper C^1 embedding theorem states that if a compact Riemannian n-manifold has a C^1 embedding into $\mathbb{R}^{n+1}$, then it has an isometric embedding into $\mathbb{R}^{n+1}$ (Kuiper 1955, Nash 1954). As a consequence of this theorem, it follows that a closed oriented Riemannian surface can be C^1 isometrically embedded into an arbitrarily small ϵ ball in Euclidean 3-space. This means that we can isometrically embed a hyperbolic disc of finite radius into an arbitrarily small ball B_ϵ in $\mathbb{R}^3$. The discrete version of this theorem was proven by Burago and Zalgaller (1995) and implies that a two-dimensional surface can be isometrically embedded into $\mathbb{R}^3$. Our crochet models with triangulations can be considered piecewise linear surfaces and can be used as visual examples of the latest theorem. In both theorems, a surface is meant to have no thickness and zero volume. In practice, our models are made of yarn, which has some thickness and some volume. It is evident that they cannot fit into a ball of arbitrarily small radius, but in the limit as the radius of the model goes to infinity, a ball of volume $V + \epsilon$ is enough to fit a model of volume V.

Next we want to understand why the models look curly in this particular way. The boundary of a hyperbolic disc D^2 can be thought of as a rope of some finite length and volume. The embedding ball in $\mathbb{R}^3$ has finite volume, and we fill it in with a thickened surface of some non-zero volume. As the radius of D^2 grows, its circumference and the volume of a tube around the boundary of D^2 grow exponentially, but the volume of the ball in $\mathbb{R}^3$ grows polynomially. This is why a model of D^2 looks less curly and more flat (or smashed) in a small neighborhood but as the radius becomes bigger, it takes up more and more space in the ball. Again, in the limit the volume of the model D^2 converges to the volume of the embedding ball as the radius of D^2 goes to infinity. The crochet model of D^2 turns into a ball quickly as the number of rows grow, and it becomes harder to complete the process further with this construction.

4. Cylinder in a Hyperbolic Space

How does a cylinder look in a hyperbolic space? Note that a *hyperbolic cylinder* is defined as a cylindrical surface in Euclidean space made of parallel lines passing through a hyperbola. There are different equivalent definitions of a cylinder in the Euclidean space, and not all of them can be used in a hyperbolic space. One that works states that the *cylinder*

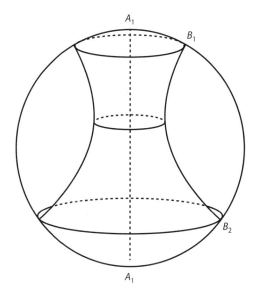

FIGURE 6. An infinite cylinder in the Poincaré ball model.

in $\mathbb{H}^3$ is a surface of revolution obtained by a rotation of a line parallel to the axis of rotation. The (infinite) cylinder cut along this geodesic line gives an (infinite) strip of H^2 with two parallel lines. They have a unique common perpendicular, which follows from the properties of hyperbolic geometry. This common perpendicular is called the *neck of a cylinder*. It is better to see a cylinder in the Poincaré model of a hyperbolic space H^3, which is a unit ball without its boundary sphere. Geodesics in this model are circular arcs perpendicular to the boundary of H^3 with chords passing through the origin.

In Figure 6, the axis of rotation A_1A_2 is fixed, and the geodesic B_1B_2 is rotated. It forms a cylinder with two ideal circles on the boundary of H^3. There is a unique plane perpendicular to both the axis A_1A_2 and the geodesic B_1B_2. Calculations in this model tell us that this plane intersects the cylinder along its neck, which looks like a circle. This is the only closed geodesic on the infinite cylinder, and it is the shortest simple closed curve on the cylinder.

A crochet model of a cylinder in Figure 7 is made out of eight equilateral triangles at each vertex. It is bounded by its neck on one side (made with yellow yarn). The other boundary is only an approximated

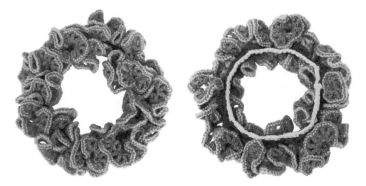

FIGURE 7. Part of an infinite cylinder bounded by its neck, top and bottom views. See also color insert.

circle; it is formed by the edges of the equilateral triangles. It is contoured with blue yarn to stretch the edges. It also makes the triangles closer to equilateral, and the model looks overall nicer.

5. Crochet and Techniques

The Gaussian curvature K of a surface at a point is defined as the product of its two principal curvatures at the given point. Although the curvature of the hyperbolic plane is a negative constant, the crochet models are only its approximations, as they have thickness along edges and are empty in the interior of the triangles. The curvature of the models is discrete at the vertices and along the edges of the triangles.

The appearance of the models depends on the yarn properties. Silkier and stretchier yarn makes models that do not hold their shapes well (they are mushy). Yarn without any elasticity, like cotton, is very hard to crochet tightly, and again it does not hold its shape as we want. Rough yarn with a little bit of stretch allows us to make firm, rigid models with beautiful patterns of embedding. Wool and acrylic yarns work well to maintain these properties. They have little elasticity, which allows us to see the local flatness of the triangles and at the same time to hold the shape of an embedded surface in $\mathbb{R}^3$.

The pattern for the seven-triangle case is written using U.S. crochet terminology (Figure 8) and is given in Figure 9. For other models, similar patterns were used. Crochet begins with a slip stitch, which is

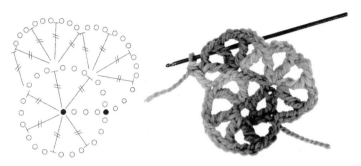

FIGURE 8. Crochet symbols (left to right): slip, chain, and triple/treble crochet stitches.

FIGURE 9. Crochet scheme for seven triangles at each vertex.

suggested to be twice as long as normal because there will be several stitches to crochet into it.

Row 1:

    Step 1: Crochet three chain stitches.

    Step 2: Crochet three chain stitches and a triple crochet stitch into the slip stitch.

    Step 3: Repeat the previous step five more times.

    Step 4: Crochet three chain stitches.

    Step 5: Connect the row with the slip stitch.

Row 2:

    Step 6: Crochet three chain stitches.

    Step 7: Crochet a triple crochet stitch into a triple crochet stitch from the previous row.

    Step 8: Repeat steps 6-7-6-7-6–7. There should be seven arcs around one loop. They represent the edges of seven triangles around a vertex.

    Step 9: Repeat steps 7–8 five more times.

    Step 10: Repeat steps 7-6-7-6-7–6.

    Step 11: Close the row with the slip stitch.

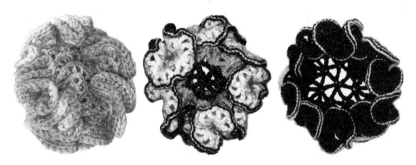

Figure 10. This is a closer look at the models near their centers. See also color insert.

Rows 3 and 4:
Repeat the pattern of Row 2 in the successive rows. The only difference is in Step 8. Sometimes there are four triple crochet stitches into a triple crochet stitch from the previous row, and sometimes there are only three of them. This number should be computed such that there are always seven arcs representing edges of triangles around a single vertex. Yarn may be changed for another color when rows change.

Row 5:
The model can be completed with slip stitches along its periphery, as in Figures 3, 7, and 10. Crochet one slip stitch into each stitch from the last row. Slip stitches can be replaced with single crochet stitches to make a thicker trimming.

6. Artist's Statement

Maria Trnkova is a Krener Assistant Professor at the University of California in Davis. She made crochet models of several hyperbolic surfaces to visualize them and to explain certain features of hyperbolic geometry. In the process of making crochet models, she used wool and acrylic yarn. She hopes her work will help others to learn hyperbolic geometry through visualization of these models. She was inspired by the crochet models of Daina Taimina, and the new models should be a useful complement to Taimina's models.

Maria's personal story of discovering triangulated crochet models of hyperbolic discs started when teaching a graduate class. She was preparing for a Riemannian geometry class and wanted to demonstrate to students how a hyperbolic disc looks and help them to visualize some of the geometric concepts. The process of crocheting a solid hyperbolic disc is exponential in time and yarn consumption. She wanted to save time and use less yarn and as a trained crocheter experimented with different stitches and techniques. Very quickly the triangular pattern came up naturally.

Acknowledgments

The author is thankful to Henry Segerman for encouragement in writing this note, to John Sullivan for asking a motivating question about why these models look this particular way, to Joel Hass for an idea of depth coloring, and to the anonymous referees for their helpful comments and suggestions.

References

Burago, Yu. D., and Zalgaller, V. A. (1995). Isometric piecewise-linear embeddings of two-dimensional manifolds with a polyhedral metric into R^3. *Algebra i Analiz*, 7(3), 76–95. (in Russian), and (1996) *St. Petersburg Mathematical Journal*, 7(3), 369–385.

Delp, K., and Thurston, W. P. (2011). Playing with surfaces: Spheres, monkey pants and zippergons. *Bridges coimbra conference proceedings.* http://www.math.cornell.edu/kdelp/surfaces.html.

Fenchel, W. (1989). *Elementary geometry in hyperbolic space.* de Gruyter stud. in math. (Vol. 11). De Gruyter.

Henderson, D. W., and Taimina, D. (2001). *Experiencing geometry: In Euclidean, spherical, and hyperbolic spaces.* Prentice Hall.

Kuiper, N. H. (1955). On C^1-isometric imbeddings I. *Indagationes Mathematicae (Proceedings)*, 58, 545–556. https://doi.org/10.1016/S1385-7258(55)50075-8.

Mann, K. (n.d.). *DIY hyperbolic geometry.* Available from the author's webpage https://e.math.cornell.edu/people/mann/papers/DIYhyp.pdf.

Nash, J. (1954). C^1 Isometric Imbeddings. *The Annals of Mathematics*, 60(3), 383–396. https://doi.org/10.2307/1969840.

Sinclair, C. (2006). Crafting geometry: What coral reefs, chaos, and the bias cut have in common. In *Craft: Transforming traditional crafts* (Vol. 1 by O'Reilly (CRT)). O'Reilly Media, Inc.

Stillwell, J. (2005). *The four pillars of geometry.* Springer-Verlag.

Taimina, D. (2018). *Crocheting adventures with hyperbolic planes: Tactile mathematics, art and craft for all to explore* (2nd ed.). CRC Press.

Taimina, D. (n.d.). *Crocheting hyperbolic planes: Daina Taimina at TEDxRiga*, video https://www.youtube.com/watch?v=D-AHvZqbMT4.

Thurston, W. P. (1997). *Three-dimensional geometry and topology.* Princeton University Press, Princeton, NJ.

Embodied Geometry in
Early Modern Theatre

Yelda Nasifoglu

> *yet how can my passions stint*
> *To see the heauens all orbicular*
> *The planets, stars & hierarchyes [a]bove*
> *All circular and Quadro quadro still[?]*
>
> ...
>
> *No no it cannot bee thee'l' neuer change*
> *Mortalls may change but they unchanged bee*
> *Ile change my forme & that's my remedy*[1]

What can a set of two-dimensional mathematical shapes say about early modern attitudes toward embodiment? It appears to offer a number of clues when placed in an academic play written during the Counter-Reformation.

Blame Not Our Author, editorially named from its opening lines, is a comedy that has survived in manuscript form in the Archives of the Venerable English College in Rome. Missing its title page, its author and precise date remain unknown, though it is presumed to have been written sometime between 1613 and 1635 by a mathematics instructor.[2] The main protagonist is Quadro, a melancholic square who wants to become a circle. With the heavens deaf to his pleas, he decides to solicit the help of Compass, "the architecte of humain wonder" (137), imagining perhaps the dignified instrument of God illustrated in the medieval *Bibles moralisées*. Instead, he gets a Daedalian trickster. Quadro's duplicitous servant Rectangulum convinces Compass that he is being played; not wanting to be outdone, the latter gives Quadro all sorts of quack medicines and purgatives to weaken his joints, binds him with hoops, and abandons him. Realizing the trickery, Quadro vows revenge on the "base and grosse mecanicke Cumpasse," (311) who has almost slain him by "false surquadrie" (312), and with this the play quickly disintegrates into geometric revenge plots, with various other shapes threatening one another with two-dimensional violence. Soon it is revealed that Line too has grievances against Compass, who has tortured him with punctures and joints, and has made him divisible in *infinitum*; and Circulum is indignant at having been manhandled by such a promiscuous instrument "turned and tossed by each Carpenter" (405). Line, in turn, has designs against Circulum, swearing to slice him into triangles, while Semicirculum warns Rhombus that Regulus will crack his legs and turn him into a trapezium. Eventually, order is restored by Regulus, the rule; punishment is meted out, and the play ends with the warning, "Let him that squars from rule and compasse bee/Vasaile to feare and base seruility" (1076–1077).

The 1613–1635 period in which the play was written predated the more radical inventions in mathematics such as the algebraization of geometry or calculus, but coincided with the debates on the nature of mathematical shapes, axioms, postulates, theorems, and proofs. The author of the play would certainly have been familiar with these. Founded in 1579, the English College was part of a network of institutions established on the continent after the Elizabethan Religious Settlement of 1559. Mostly run by the Jesuits, the College took on students for the study of philosophy and theology, preparing them for their mission to convert England back to Catholicism. The author would have attended lectures at the Collegio Romano, and judging from the abundant references in the play, had access to the authoritative commentaries on Euclid by the famed Jesuit professor of mathematics, Christoph Clavius (1538–1612).

One of the debates performed in the play is on the ontology of mathematical shapes. In Euclidean geometry, definitions played the key role (a triangle was defined by its three sides, a square by its four equal sides, etc.), and thus any alteration of a form would irreversibly change its identity at a fundamental level, a cause of serious concern for the characters. For instance, reveling in the excesses of the Carnivale, gluttonous Line is concerned he has eaten so much that he is about to add some breadth to his length, pushing himself out of his own Euclidean definition. And would Quadro, defined by and named after its very four sides, still be *quadro* if indeed Compass helped it become a circle? Such a metamorphosis would alter the definition of a quadrangle: Rectangulum is anxious that should his master reach his goal, all their square plates would turn into round dishes and their four-cornered meat pies into Norfolk dumplings, or worse, into round fruits. This close alignment between form and its definition makes the threats of violence, of cutting shapes or breaking their legs, into threats of annihilation. However, apart from a bruise or two, no mathematical shape is seriously hurt during the course of the play, nor is Quadro converted into a circle.

Of course in this particular case, Compass could not have helped Quadro perfect its shape even if he had wanted to. Compass and rule, the mathematical instruments featured in the play, were often factored into Euclidean constructions; for example, Book 1, Postulate 1 explained that a straight line from any point to any point can be drawn with a rule, and Proposition 1 described how to draw an equilateral triangle with the use of a rule and compass. As powerful as these instruments may have been in terms of constructing mathematical bodies, they were ineffective when it came to the squaring of the circle, one of the three classical problems that cannot be solved using a rule and compass. The problem refers to defining a square with the exact area of a circle or vice versa, but π, being an irrational number that cannot be expressed as a fraction (a fact that eluded mathematicians until the eighteenth century), this is not possible to accomplish using Euclidean geometry.

The "false surgery" Compass performs on Quadro, loosening his joints and paring off his corners, seemingly prepares him for Archimedes' method of approximation of π by breaking the sides of a polygon inscribed within a circle into smaller and smaller units. Yet we soon find another operation being performed on stage here, one that takes us outside of the realm of pure mathematics. With Quadro so eager to be rounded off into a *punctum*, Compass easily lures him into an instrument he calls the "Squarenighers daughter," or Scavenger's daughter, named after its inventor Skevington, a lieutenant of the Tower of London during the reign of Henry VIII. Rather than stretching on the rack, this device worked by severely contracting the body into a circle: the prisoner would be forced to kneel on the floor, the hoop would be placed over him and compressed until the abdomen, thighs, and lower legs were pressed into one another. Thomas Cottam (1549–1582), who had taught at the English Catholic College in Douai, France, was one of the victims of this instrument.

For a play featuring abstract mathematical shapes, there is an unexpected amount of violence, no doubt a reflection of the world surrounding the stage. The main goal of the College, of reconverting England to Catholicism, was considered in England an act of sedition punishable by the cruel mathematics of hanging, drawing, and quartering, and indeed between 1581 and 1679, the College saw 41 of its alumni martyred for their attempts, 32 of these before 1616.[3] Rather than a fate to be avoided, martyrdom was celebrated if not glorified; in 1583, Niccolò Circignani (1520–1596) was commissioned to paint a cycle of 34 frescoes in the College chapel, featuring images of hangings and violent mutilations of victims from early Christianity to the English Mission, depicting the demise of at least one alumnus of the College.[4] The students were encouraged to contemplate these images and even self-flagellate, perhaps in preparation for their possible fate in England. Martyrdom was a necessity for the survival of the church; the spectacle of public torture and execution of Catholic priests was expected to convert viewers impressed with the courage and intense faith of the victims. There are additional martyrology references in the play, such as the surprisingly irreverent ones by Line and Rectangulum regarding the hanging of "Papists." Furthermore, the "presse" Rectangulum tricks some of the other mathematical forms to hide in may well have been the type of large grape or olive press used in the martyrdom of Saint Jonas as illustrated in Antonio Gallonio's *Trattato de gli instrumenti di martirio*, "Treatise of the instruments of martyrdom" (1591 in Italian, 1594 in Latin).

Though he ultimately fails in his quest, Quadro longs to attain perfection by changing his form into a circle, willing to endure even torture in the process. And there are stories about self-discipline, of following strict daily regimens, being lauded at the College, hinting at a perceived relationship between changes to the body and their positive effect on the soul. Yet there is an underlying anxiety about such a close relationship, especially when the transformations are not self-inflicted or desired: if a rhombus with broken legs is a trapezium, what happens to the martyr whose body is drastically reconfigured in the hands of his or her torturer? No clear answer to this is offered in the play, of course, but the theatres of cruelty illustrated by the martyrdom cycle in the chapel or in treatises like Gallonio's seem to downplay the effectiveness of worldly instruments against the fortitude of their victims; just as they are unable to square the circle, despite all the pain they can inflict, instruments appear to be powerless against the resolve of the faithful.

Acknowledgments

Many thanks to Simon Ditchfield for his comments and suggestions on an earlier version of this text. This piece was a short reflection on the idea of embodiment for an Oxford Philosophical Concepts volume on the subject edited by Justin E. H. Smith.

Notes

1. The primary source is Gossett 1983, 94–132 at 96; lines 30–33, 38–40; hereafter only line numbers will be indicated.
2. Gossett 1973. For previous scholarship on this play, see also Mazzio 2004.
3. Sources on the history of the College include Foley 1880 and Williams 2008.
4. On Circignani's martyrology cycle, see Williams 2005.

Bibliography

PRIMARY SOURCES

Gossett, Suzanne, Ed. "Blame Not Our Author." In *Collections, Volume XII*. Oxford, U.K.: The Malone Society, 1983, 94–132.

SECONDARY SOURCES

Foley, Henry. *Records of the English Province of the Society of Jesus, Vol. VI*. London: Burns and Oates, 1880.

Gossett, Suzanne. "Drama in the English College, Rome, 1591–1660." *English Literary Renaissance* 3, no. 1 (1973): 60–93.

Gossett, Suzanne. "Introduction [to *Blame Not Our Author*]." In *Collections, Volume XII*, Suzanne Gossett, Ed., Oxford, U.K.: The Malone Society, 1983, 83–93.

Mazzio, Carla. "The Three-Dimensional Self: Geometry, Melancholy, Drama." In *Arts of Calculation: Quantifying Thought in Early Modern Europe*, David Glimp and Michelle R. Warren, Eds., New York: Palgrave Macmillan, 2004, 39–65.

Williams, Michael E. *The Venerable English College, Rome: A History*, 2nd ed. Herefordshire, U.K.: Gracewing, 2008.

Williams, Richard L. "Ancient Bodies and Contested Identities in the English College Martyrdom Cycle, Rome." In *Roman Bodies: Antiquity to the Eighteenth Century*, Andrew Hopkins and Maria Wyke, Eds., London: The British School at Rome, 2005, 185–200.

Modeling Dynamical Systems for 3D Printing

STEPHEN K. LUCAS, EVELYN SANDER,
AND LAURA TAALMAN

1. Motivation

As 3D printers become increasingly common in science and engineering, they are also making their way into mathematics departments. One potential use is in visualizing dynamical systems. These mathematical structures illustrate some of the important progress made in roughly the past 60 years regarding approaches to understanding the role of dynamical models in scientific research. In addition to the illustration of principles, we are in no small part motivated by the beautiful three-dimensional structures of strange chaotic attractors, such as can be seen in [5]. *See*, for example, the Langford attractor in Figure 1. The aim of this paper is to describe how to 3D print actual physical models of such dynamical structures.

In many cases, it is possible to use a black-box differential equation solver to create printable objects. However, in some cases, a black-box method is inadequate. We have therefore developed a mixed curvature method to make printing possible in these cases. We mostly restrict our discussion to solutions of differential equations, but we end with a description of how 3D printing can be applied in the context of more general dynamical structures for iterated maps and ordinary and partial differential equations.

Our paper proceeds as follows: In Section 2, we give a straightforward method for generating 3D printable chaotic attractors using built-in commands in Mathematica. In Section 3, we describe a new method for creating more visually accurate 3D printable chaotic attractors. This mixed curvature method uses a combination of MATLAB and

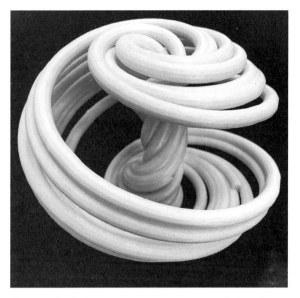

FIGURE 1. The Langford chaotic attractor in Equation (8) (often errone-ously referred to as the Aizawa attractor), modeled in Mathematica and 3D printed in fused deposition modeling (FDM) polylactic acid (PLA). See also color insert.

the cost-free software OpenSCAD [11], designed specifically for 3D print design. In Section 4, we describe our future directions in creating printable chaotic and dynamical objects. In Appendix A, we provide all equations and initial conditions of the objects shown in the figures. In Appendix B, we give a brief introduction to the nuts and bolts of 3D printing so that readers will be able to use the methods provided in this paper to create their own printed attractors. All code used is available for download at [14].

2. Straightforward Approach

Strange chaotic attractors have interesting dynamical properties, such as the plethora of periodic orbits (in fact, infinitely many) and sensitive dependence on initial conditions, meaning that no matter how close solutions start, they will diverge over time [20]. These properties are not visible when looking at the printed attractor. Rather, one sees the strangeness, or fractal, structure of the attractor. Any cross section of

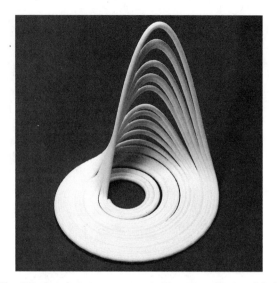

FIGURE 2. The Rössler chaotic attractor in Equation (5), modeled in
MATLAB and OpenSCAD and 3D printed in selective laser sintering (SLS)
nylon. Photo credit: Edmund Harriss. See also color insert.

the attractors shown here consists of a Cantor set: The fractal dimension of the attractors is between 1 and 2. This can be seen particularly
well in the Rössler attractor in Figure 2, the Lorenz attractor in Figure
3, the Rucklidge attractor in Figure 4, and the Anishchenko attractor
in Figure 5.

From the practical point of view of generating chaotic attractors, the
most useful properties are the existence of a dense orbit and the fact
that the set is attracting. The first property implies that we can create
an arbitrarily good approximation of the chaotic attractor with a single
solution starting within the attractor. The second property implies that
we do not even need to start with a solution within the attractor. In
particular, as long as we start our solution close enough to the attractor, the solution will converge to the attractor as time increases. Based
on these principles, in this section we describe how to generate a 3D
printable solution to a given differential equation starting at a specified
initial point.

Mathematical software packages such as Mathematica and MATLAB
have built-in routines for solving systems of first-order ordinary differential equations. They allow for the creation of solutions to differential

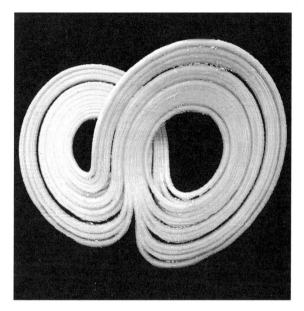

FIGURE 3. The Lorenz chaotic attractor in Equation (4), modeled in Mathematica and 3D printed in FDM PLA. See also color insert.

FIGURE 4. The Rucklidge chaotic attractor in Equation (10), modeled in MATLAB and OpenSCAD and 3D printed in SLS nylon. Photo credit: Edmund Harriss. See also color insert.

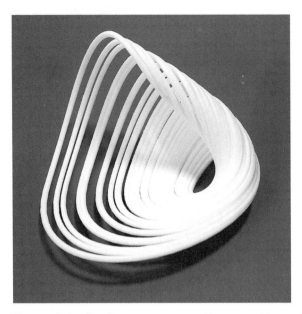

FIGURE 5. The Anishchenko chaotic attractor in Equation (11), modeled in MATLAB and OpenSCAD and 3D printed in SLS nylon. Photo credit: Edmund Harriss. See also color insert.

equations without prior knowledge of numerical methods and relatively minimal knowledge of the theory of differential equations. Here we present a method using the NDSolve command in Mathematica. The choice of Mathematica over MATLAB is based on the fact that the package has better developed methods for creating 3D printable objects, meaning that it is possible to generate data for and create a printable (STL format) file within a single software package. This has been successfully used in a class consisting of undergraduate students who had only seen elementary differential equations and had never seen any numerical methods for solving differential equations.

The Mathematica command NDSolve is proprietary and to some extent a black box, though it is possible to change some of the default settings. In particular, the user inputs an equation of the form

$$\dot{\mathbf{x}} = f(t, \mathbf{x}), \ \mathbf{x}(t_0) = \mathbf{x}_0$$

where $\mathbf{x}(t)$, $\mathbf{x}_0(t) \in R^n$, and $f: R \times R^n \to R^n$. Here and in all subsequent equations, we use the notation $\dot{\mathbf{x}}$ to represent the derivative of $\mathbf{x}$. In most of the cases we consider $n = 3$, since that is the dimension of our printing space. The input of the method includes the form of the equation, the initial condition, and the minimum $t_0 = 0$ and maximum T time values for which a solution should be calculated. The method is an adaptive Runge-Kutta one that makes an approximation of the error and correspondingly adjusts the size of the time step. It outputs an interpolation for $\mathbf{x}(t)$ on a discrete but unevenly spaced set of $t \in [0, T]$. The method determines the spacing of the t values such that the approximate solution achieves a desired accuracy. That is, at each t if we denote by $\mathbf{x}_{\text{exact}}(t)$ and $\mathbf{x}_{\text{approx}}(t)$ the exact and approximate solutions, respectively, then the method computes approximations with prescribed small values of the absolute error $|\mathbf{x}_{\text{exact}}(t) - \mathbf{x}_{\text{approx}}(t)|$ and relative error $|\mathbf{x}_{\text{exact}}(t) - \mathbf{x}_{\text{approx}}(t)| / |\mathbf{x}_{\text{exact}}(t)|$. Note that $\mathbf{x}_{\text{exact}}$ is unknown, but the method still is able to say with some confidence that the approximate solution is sufficiently close to the exact one in terms of both of these measures of error. When these desired accuracy requirements are not set by the user, they remain at default values internal within the software package.

After generating an approximation in the form of an interpolation for $\mathbf{x}(t)$ for $t \in [0, T]$, we can then use the Mathematica command `ParametricPlot3D` to create a piecewise smooth curve approximating the exact solution. By using the `PlotPoints` option with a large value such as 100, we are able to control the level of smoothness of the curve. At this stage, the curve is not a solid printable object since it is infinitely thin. We use the `Tube` graphing option to create a tube of a specified thickness around the curve, thereby making it into a solid and therefore printable object. This solid object can be saved in printable STL format. *See* Appendix B for more information on how to proceed from an STL file to a physical object.

While our primary goal is to create chaotic attractors for ordinary differential equations, with minimal extra effort the same commands in Mathematica can be used to generate solutions to delay differential equations. Figure 8 depicts a chaotic attractor for the Mackey-Glass delay differential equation. That is, the derivative $\dot{\mathbf{x}}(t)$ depends not only on t and $\mathbf{x}(t)$, but also on the value of $\mathbf{x}$ at a previous time $t - \tau$ for $\tau > 0$. In addition to chaotic attractors, the method works equally well for generating other types of solutions to differential equations. For example, Figure 7 shows a quasiperiodic torus solution to the Hénon-Heiles equation.

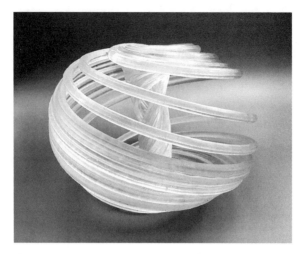

FIGURE 6. The Langford chaotic attractor in Equation (8), modeled in MATLAB and OpenSCAD and 3D printed in stereolithography (SLA) resin. See also color insert.

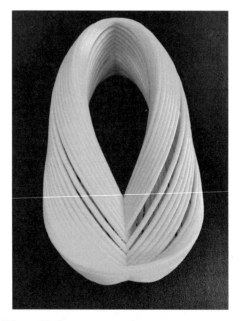

FIGURE 7. The Hénon-Heiles quasiperiodic set in Equation (9), projected to three dimensions, modeled in Mathematica and 3D printed in FDM PLA. See also color insert.

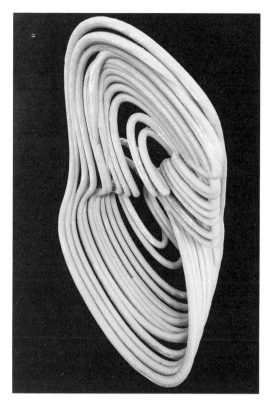

FIGURE 8. The Mackey-Glass attractor in Equation (12), projected to three dimensions, modeled in Mathematica and 3D printed in FDM PLA. See also color insert.

3. Mixed Curvature Method

In this section, we present a new mixed curvature method for preparing a 3D printable differential equation solution. The method produces a file with a smaller number of data points, but still with the same or in some cases greater visual accuracy. In many cases, the straightforward approach described in Section 2 is completely sufficient for generating a visibly excellent approximation of a chaotic attractor. For example, Figures 9 and 10 are both equally acceptable 3D printed representations of the Arneodo attractor, where Figure 9 was created using the straightforward method and Figure 10 used the mixed curvature approach. The same applies to the Langford attractor shown in Figures 1 and 6.

FIGURE 9. The Arneodo chaotic attractor in Equation (7), modeled using the straightforward method in Mathematica and 3D printed in FDM PLA. See also color insert.

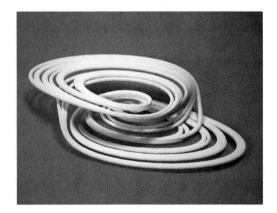

FIGURE 10. The Arneodo chaotic attractor in Equation (7), modeled using the mixed curvature method in MATLAB and OpenSCAD, and 3D printed in SLS nylon. Photo credit: Edmund Harriss. See also color insert.

There are, however, cases where the straightforward approach does not lead to a visually accurate object, usually when the curvature of the solution curve varies substantially along its length. The straightforward approach tends to produce curves that lack the required smoothness in the high-curvature regions, since the data distribution is not considering the shape of the object that we wish to eventually print. Attempts to fix this by increasing the requested density of

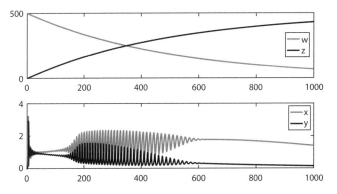

FIGURE 11. Solutions to the autocatalytic system in Equation (1) using
ode45 as a function of time. See also color insert.

points lead to a vast data set that is difficult for printers to work with,
and they contain superfluous data in the low-curvature areas of the
curve. An example where this behavior is particularly noticeable is
when attempting to visualize the solution to the autocatalytic chemical
reaction system [9]:

$$\dot{w} = -0.002w$$
$$\dot{x} = 0.002w - 0.08x - xy^2 \qquad (1)$$
$$\dot{y} = 0.08x + xy^2 - y$$
$$\dot{z} = y$$

with $w(0) = 500$, $x(0) = 0$, $y(0) = 0$, and $z(0) = 0$. This system does not
have an analytic solution (apart from $w = 500e^{-0.002t}$), but it can be approximated to prescribed accuracy using a standard numerical solver,
such as NDSolve in Mathematica, or ode45 in MATLAB. Figure 11
shows the solutions produced using ode45 to the autocatalytic system
with the default settings on the interval $t \in [0, 1,000]$. Two plots are
shown due to the very different size scales involved in the solutions; the
ode45 approximation contains 2,093 data points.

However, a more intuitive visualization is to plot a projection in
three-dimensional space of the position $(x(t), y(t), z(t))$. Dividing z by
100 and making the axes of equal length, we get the picture in Figure
14. Note that if we did not scale the z variable, then the curve would be
extremely long and thin, and inappropriate for 3D printing. While not
as compact as the chaotic attractors, this is still an interesting object for

3D printing, as shown in Figure 15. In all that follows, we scale the *z* axis for the autocatalytic system.

3.1. DEFAULT POINT PLACEMENT

Unfortunately, the data that produced Figures 11 and 14 are not particularly useful when developing a structure for 3D printing. If one is not using the built-in Mathematica routines, the standard approach is to take a set of data points along the curve, join them by straight-line segments, then expand them into tubes with some given cross section of a pleasing radius. This step is identical to the use of the Mathematica `Tube` command used in the straightforward approach. Having the points placed so that the joins between the tubes do not form obvious corners leads to a structure that is more pleasing to the eye. Unfortunately, the typical data output from a numerical solver is not of this form. An adaptive solver like `ode45` tries to use the minimum number of points in time to satisfy some requested error bound over the interval of interest. Even though solutions between data points can be interpolated, the velocity along a curve usually varies, and straight-line tubes are of wildly varying lengths. In this particular example, the default `ode45` output looks quite poor in the middle parts of the solution (*see* Figure 13), where the *x* and *y* components are oscillating.

3.2. EQUALLY SPACED POINTS IN SPACE

As a first attempt beyond default output data points, we require the data points that define the tube endpoints to be equally spaced along the curve in the three-dimensional phase space. This can be done most efficiently by adding an additional differential equation to the original set that gives the length of the curve so far as a function of time. Given that the arc length of the curve defined by $(x(t), y(t), z(t))$, $0 \le t \le T$ is

$$s(t) = \int_0^t \sqrt{\dot{x}(u) + \dot{y}(u) + \dot{z}(u)} \, du$$

Taking the derivative gives us

$$\dot{s}(t) = \sqrt{\dot{x}(t) + \dot{y}(t) + \dot{z}(t)} \text{ with } s(0) = 0$$

Formulas for $\dot{x}$, $\dot{y}$, and $\dot{z}$ are given in the original system of differential equations. For example, Figure 12 shows the length of the curve for

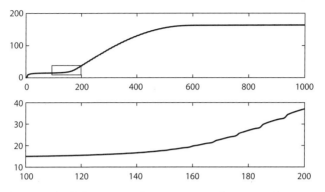

FIGURE 12. Length of the autocatalytic curve as a function of time (scaled in the *z* direction). The lower graph shows a closer view of the boxed region outlined in the upper graph.

the autocatalytic system, with *z* divided by 100, as a function of time. Clearly, the velocity is not constant along this curve, which has total length 163.45, and the default output produces line segments that vary in length from 5.02×10^{-5} to 4.53.

To produce data points equally spaced in terms of distance along a curve on some time interval [0, *T*], since we have the distance function *s*(*t*), we want data at times $\{t_i\}_{i=0}^{m}$ that satisfy

$$s(t) - \frac{is(T)}{m} = 0 \text{ for } \quad i - 0, 1, \dots, m$$

where $t_0 = 0$ and $t_m = T$. An easy way to solve for the t_i values assumes we know *s*(*t*) for $0 \le t \le T$, and then use the secant method with the initial guesses t_i and $t_i + 10^{-3}$ when trying to find t_{i+1}. While MATLAB's ode45 outputs the solutions at particular points, it allows for interpolation between these data points, so we can find the length of the curve at any time. Since *s* is an increasing function, the secant method is a robust way of finding these data points.

For example, Figure 13 shows the three-dimensional curve for the autocatalytic problem with 3,000 equally spaced points in terms of arc length. It is immediately obvious that this is not a particularly good representation. Even with more data points than output from ode45, there are places along the curve where it is clearly not smooth, particularly in the lower section after the initial spiral inward.

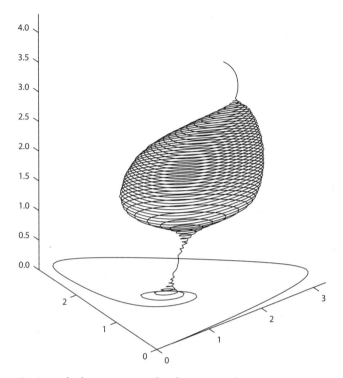

FIGURE 13. Straight-line segments for the autocatalytic system in Equation (1) scaled in the z direction with 3,000 equally spaced (in terms of arc length) points.

3.3. BOUNDING CURVATURE

The problem with the equally spaced approach is that regions of the curve with high curvature lead to straight-line segments with a noticeable corner between them. An alternative approach, then, is to choose data points so that the maximum distance from the actual curve to the approximating straight-line segment has an upper bound. The radius of curvature of the curve could be calculated along the curve with some additional analysis, and keeping the curve close to the line segments is equivalent to bounding the angle of the sector of the osculating circle that the curve moves through between data points.

While it is theoretically possible to calculate the radius of curvature along the curve and choose an interval length associated with the smallest radius of curvature on the interval, it turns out to be a challenging numerical exercise not worth detailing here. Even worse, we found that simply using the radius of curvature was also a poor choice. Choosing interval lengths to place a sufficient number of points in regions of high curvature means far too many points were placed in regions where the curvature is low.

3.4. MIXED CURVATURE METHOD

Ideally, we wish to have a combination approach, with some upper bound on the arc length between points when the curvature is low, then placing more points where the curvature is high. While sophisticated analysis is possible, we suggest the following simple approach that works well in every situation we have considered. Start with the equally spaced in space approach, giving some initial length of line segments that initially appears reasonable when the curvature is not high. Then, if either of the angles at the joins of the current line segment with its neighbors is too small, less than some angle $(180 - d)°$, bisect the line segment by adding a data point at the middle of the arc length curve, splitting the original line segment into two equal-length pieces, thus increasing the density of data points in regions of high curvature. If any line segment has been split, repeat a full pass over the current set of line segments, splitting as necessary, until eventually all of the angles between line segments are sufficiently close to 180°.

For the autocatalytic example with time interval [0, 1,000], we chose to start with 1,000 equally spaced points and angle bound $d = 10$. The choice of 1,000 points made the relatively straight parts of the curve look good: Even small angles are more noticeable between straight line segments when they are long. After six passes subdividing problematic intervals, we ended up with 4,012 data points and the picture in Figure 14.

Once data points for a curve have been established by the mixed curvature method, we need to transform from line segments to tubes and form an STL format file for 3D printing. Our experience thus far is that this process is not particularly straightforward or successful within MATLAB, and so we turned to using the application OpenSCAD [11] to take as input these data, produce a tubular structure, then form an STL file. Figures 2, 4, 5, 6, 10, 15, and 21 were 3D printed using the mixed curvature method.

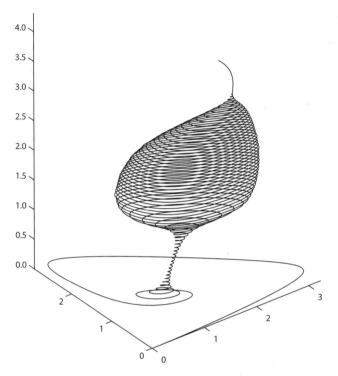

FIGURE 14. Straight-line segments for the autocatalytic system scaled in the z direction with initially 1,000 points and minimum angle 170°.

FIGURE 15. Autocatalytic system in Equation (1) with scaled z variable modeled with the mixed curvature method and 3D printed in SLS nylon.

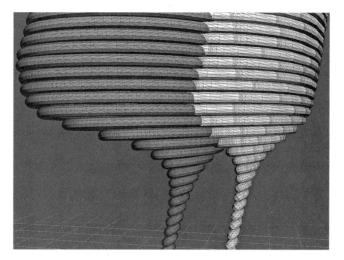

FIGURE 16. Mesh comparison of straightforward (light model) and mixed curvature (dark model) methods for the scaled autocatalytic system. *Note* the higher accuracy and more appropriate mesh variation in the mixed curvature model.

It is appropriate at this time to compare the STL files formed by the straightforward and mixed curvature approaches. As we have stated, for most of the chaotic attractors we have printed and shown previously, it is difficult to tell which approach has been used. Most of the time, the straightforward approach is by far the simplest and best. But Figure 16 places the mesh formed by the two methods in the same visualization space. It is obvious that the mixed curvature model provides a more pleasing object to print.

4. Future Directions

While it is not hard to find examples of 3D printed objects related to both multivariable calculus and geometry, there is relatively less material available on the topic of 3D printing in dynamical systems. The examples in this article are only one step in filling such a gap, and there is plenty of room for future work in this direction. In particular, there are many dynamical objects beyond solutions to and attractors of differential equations that are well suited for 3D printing. We include a few examples of possible future directions below.

The methods described in this paper can be used to consider differential equations that vary with respect to a parameter, and they can be used to create a series of attractors in three dimensions as a parameter varies. This allows for the visualization of bifurcations in attracting sets. An example of this appeared in [8].

Attractors of three-dimensional iterated maps are equally well suited to 3D printing as those of differential equations. However, since the orbit of an iterated map consists of a set of disconnected points that only form a connected set when combined, it is more difficult to create a printable mesh. The most obvious method of combining such points is to create a small sphere at each iteration and combine these overlapping spheres to create an object. While this is doable, it results in an object with an extremely large file size. In addition, overlapping objects can cause mesh errors because it is not clear what to interpret as the boundary of the objects, and this is only exacerbated by the fractal nature of a strange attractor. These can often be fixed using a mesh repair program, but this is an extra and not always successful step. A more in-depth mathematical solution to the creation of an attractor of an iterated map is to use the structure of the chaotic attractors.

An attractor of a three-dimensional iterated map almost always falls into one of two categories, a multiply folded one-dimensional curve or a two-dimensional surface, described in [1] as the *spaghetti-lasagna dichotomy*. In the spaghetti case, the attractor is typically the closure of a branch of a one-dimensional unstable manifold of a fixed point or periodic point of the map. The advantage of using this characterization is that the unstable manifold is a connected curve, and therefore there are no longer the same meshing issues as with small spheres. Figure 17 shows $(x, y, h_2^{-1}(x, y))$ for an attractor of the Hénon

$$h(x, y) = (1.4 - x^2 - 0.3y, x) \qquad (2)$$

created using a branch of the unstable manifold. The attractor lies in two dimensions, but we have included a preimage for a delay embedded version, making the graph three dimensional. The creation of attractors that arise as the closure of a curve is rather straightforward, though it involves more mathematical discussion than the differential equation case. However, the creation of a lasagna-type attractor surface is quite involved and better done using previously written software packages.

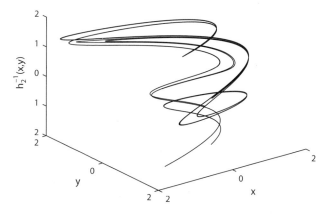

FIGURE 17. A delay embedded Hénon attractor in Equation (2) created using a branch of the unstable manifold.

For an idea of the difficulties involved, the creation of a crocheted Lorenz unstable manifold is described in [17].

For a two-dimensional dynamical system that varies with respect to a parameter, each attractor or other dynamical object is contained in a plane. Therefore, we can explore variation with respect to a parameter in the scope of a single 3D print. Namely, each dynamical object occurring at a fixed parameter occurs at a single slice, and together the slices form a 3D object. For example, the Chirikov standard map in Equation (13) in Appendix A is an area-preserving map on the unit square that varies with respect to a parameter k. At each fixed k value between $k = 0$ and $k \approx 0.971635406$, there is an invariant curve consisting of the closure of a single periodic orbit with the property that the rotation number (average of $f(x) - x$ along the orbit) is equal to the golden mean $(\sqrt{5} - 1)/2$. Rather than being chaotic, these invariant curves are *quasiperiodic*. Figure 18 shows how the curves vary with respect to k for $0 < k < 0.9716$. At each fixed k, we use a root-finding method to find the curve with the given rotation number. Figure 19 is a photo of a 3D printed model of these curves.

We also note that iterated maps can be made higher dimensional by including point density information using a binning approach, where we count how often a map visits regions in some mesh. For example, Figure 20 shows the iterated logistic map

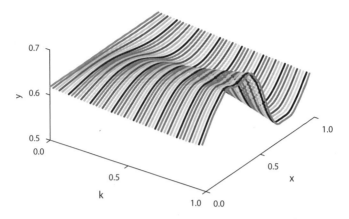

FIGURE 18. For the Chirikov standard map in Equation (13), the quasiperiodic curve with golden mean rotation number varying with respect to parameter *k*. See also color insert.

FIGURE 19. 3D printed model of the curve in Figure 18 in PLA with colored stripes made possible by using a Palette filament splicer. See also color insert.

$$x_{n+1} = rx_n(1 - x_n) \qquad (3)$$

where the height is a measure of how likely it is for the map to visit a particular region as we vary *r*.

Finally, parabolic partial differential equations can also be viewed as dynamical systems for solutions of the form $u(t, x)$, and at fixed *t* values,

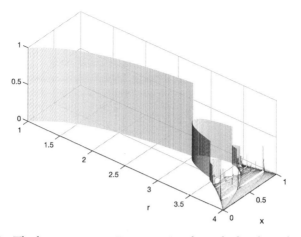

FIGURE 20. The logistic map in Equation (3) where the height is the probability of visiting a particular location as r varies.

the solution u can have rich structure. These provide a rich set of examples for creation of 3D prints. An example of a 3D print of a spinodal decomposition in the Cahn-Hilliard equation is in [21].

A. List of Dynamical Systems

We have tested the standard algorithm on the following examples, chosen both for beauty and for importance. Many of these examples can be found in [1] and [16].

- Lorenz attractor [13]:

$$\begin{aligned}
\dot{x} &= \sigma(y - x) \\
\dot{y} &= x(\rho - z) - y \\
\dot{z} &= xy - \beta z
\end{aligned} \tag{4}$$

The printed object was creating using $\sigma = 3$, $\beta = 1$, and $\rho = 28$, and initial conditions $x_0 = 0$, $y_0 = 1$, $z_0 = 0$. The Lorenz system is the first chaotic attractor within the scientific community. It is an atmospheric model created to understand unpredictability of linear models in weather prediction. *See* Chapter 9 of [1] for a detailed discussion of the history of the model.

- Rössler attractor [18]:

$$\dot{x} = -y - z$$
$$\dot{y} = x + ay \qquad (5)$$
$$\dot{z} = b + z(x - c)$$

The printed object was creating using $a = 0.1$, $b = 0.1$, $c = 18$, and initial conditions $x_0 = 0$, $y_0 = 1$, $z_0 = 0$. This system was created to show that chaos could occur in systems that were even simpler than the Lorenz equations.

- Chen double scroll attractor [6]:

$$\dot{x} = a(y - x)$$
$$\dot{y} = (c - a)x - xz + cy \qquad (6)$$
$$\dot{z} = xy - bz$$

The printed object was creating using $a = 40$, $b = 3$, $c = 28$, and initial conditions $x_0 = -0.1$, $y_0 = 0.5$, $z_0 = -0.6$. This system was created to exhibit properties of both Lorenz and Rössler attractors.

- Arneodo attractor [4]:

$$\dot{x} = y$$
$$\dot{y} = z \qquad (7)$$
$$\dot{z} = -\alpha x - \beta y - z + \delta x^3$$

The printed object was creating using $\alpha = -5.5$, $\beta = 3.5$, $\delta = -1$, and initial conditions $x_0 = 0.2$, $y_0 = 0.2$, $z_0 = -0.75$. This attractor was developed to illustrate chaos in a physical system near a triple instability.

- Langford attractor [12]:

$$\dot{x} = (z - b)x - dy$$
$$\dot{y} = dx + (z - b)y \qquad (8)$$
$$\dot{z} = c + az - z^3/3 - (x^2 + y^2)(1 + ez) + f z x^3$$

The printed object was creating using $a = 0.95$, $b = 0.7$, $c = 0.6$, $d = 3.5$, $e = 0.25$, $f = 0.1$, and initial conditions $x_0 = 0.1$, $y_0 = 1$, $z_0 = 0$. Note that while this attractor is commonly named after Yoji Aizawa, it cannot be found within his published work, and was in fact developed by Langford [12]. He was investigating models of chaotic behavior moving upon a torus.

- Hénon-Heiles invariant torus [10]:

$$\dot{x} = z$$
$$\dot{y} = w \qquad (9)$$
$$\dot{z} = -x - 2xy$$
$$\dot{w} = -y - x^2 + y^2$$

The printed object was creating using initial conditions $x_0 = 0$, $y_0 = 0$, $z_0 = 0.35$, $w_0 = -0.3$. The total time length is 200. This system models the motion of individual stars as affected by the rest of a galaxy. Unlike the other models in this paper, this is a four-dimensional example. We project and only plot (x, y, z). The system is Hamiltonian, so the chaotic solutions are not strange attractors and thus do not make very pretty prints. We have chosen an initial condition corresponding to a quasiperiodic solution (it has energy approximately 0.10625), which is a topological torus in four dimensions.

- Rucklidge attractor [19]:

$$\dot{x} = \kappa x - \lambda y - yz$$
$$\dot{y} = x \qquad (10)$$
$$\dot{z} = -z + y^2$$

The printed object was creating using $\kappa = -2$, $\lambda = -6.7$, and initial conditions $x_0 = 1$, $y_0 = 0$, $z_0 = 4.5$. This system came about when modeling two-dimensional convection in a fluid layer rotating uniformly about a vertical axis.

- Anishchenko-Astakhov attractor [2]:

$$\dot{x} = \mu x + y - xz$$
$$\dot{y} = -x \qquad (11)$$
$$\dot{z} = -\eta z + \eta H(x)x^2$$

The printed object was creating using $\mu = 1.2$, $\eta = 0.5$, $H(x)$ is the Heaviside function, and initial conditions $x_0 = -0.1$, $y_0 = 0.5$, $z_0 = -0.6$. This system was proposed in the study of nonlinear oscillators. The original paper is in Russian, but a description appears in English in [3].

- Mackey-Glass attractor [15]:

$$\dot{x}(t) = \beta x(t - \tau)/(1 + x(t - \tau)^n) - \gamma x(t) \qquad (12)$$

The printed object was creating using $\gamma = 1$, $\beta = 2$, $\tau = 2$, and $n = 9.65$, and initial condition $x(t) = t^2$ for $-\tau < t < 0$. This equation models dynamical diseases including respiratory disorders, such as irregular breathing and apnea, and hematologic disorders, such as chronic myelogenous leukemia, in which blood cell counts oscillate rather than staying constant. Unlike other examples, this produces a strange chaotic attractor for a delay differential equation. We plot the solution in three dimensions by projecting to $(x(t), x(t - \tau), \dot{x}(t - \tau))$. Both Mathematica and MATLAB have built-in delay equation solvers, making it possible to use a black-box code for solving this equation.

- Chirikov standard map [7]:

$$x_{t+1} = x_t + y_{t+1} \bmod 1$$
$$y_{t+1} = y_t - \frac{k}{2\pi} \sin(2\pi x_t) \qquad (13)$$

This iterated map is a well-known example in the study of area-preserving maps and Kolmogorov-Arnold-Moser theory. When $k = 0$, all points lie on "rotational" invariant circles, which are graphs of x as a function of y. For each fixed rotation number $\omega \in (0, 1)$, rotational invariant circles exist and vary smoothly for a parameter interval $(0, k_\omega)$. The largest value of k_ω occurs when ω is equal to the golden mean (or its inverse).

B. 3D Printing Notes

The physical models photographed in this paper were 3D printed using a variety of methods, which we outline here so that others can use these techniques to print their own models.

The 3D models shown in Figures 1, 3, 7, 8, and 9 were 3D printed with desktop fused-deposition modeling (FDM) machines in polylactic acid (PLA) plastic. When printing with FDM, one layer of plastic is drawn out at a time, each supporting the next. If your model has overhangs (as is certainly the case for these attractors), then you also

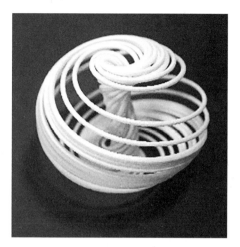

FIGURE 21. The Langford chaotic attractor in Equation (8). Modeled in MATLAB and OpenSCAD and 3D printed in SLS nylon. Photo credit: Edmund Harriss. See also color insert.

have to include extensive support material as part of your print. This support material can be removed after printing, but it leaves marks and can damage or break the model during cleanup. For this reason, models printed with FDM can have a rough surface and in addition need to have thicker path diameter for strength, so less detail is possible.

The model in Figure 6 was printed on a desktop resin 3D printer using stereolithography (SLA) technology. This method uses a laser to harden liquid resin one thin layer at a time, with the model developing upside down while attached from a build plate that dips into a pool of resin. Models printed with SLA can in general be very delicate, and the sweeping curves of attractors do tend to break during support removal and cleanup, so a thick path diameter is recommended with this method as well. SLA printing also requires washing with isopropyl alcohol and curing with sunlight or a UV light. The final printed models have a high-quality finish.

The models shown in Figures 2, 4, 5, 10, 15, and 21 were printed by the service bureau Shapeways in nylon plastic, using selective laser sintering (SLS). With this technology, the model is created by depositing very thin layers of powder, which are solidified by a laser in the spots that intersect the design. At the end of printing, the model is completely encased and supported by loose powder, so there are no supports to remove. Very thin and detailed models can be printed successfully with this method. Compare the detail in the SLS model of the Langford attractor shown Figure 21 with the coarser models of about the same overall size printed in FDM (Figure 1) and SLA (Figure 6). SLS printing is a particularly good option for art/display-quality models and for those without their own 3D printers in house.

Finally, the model in Figure 19 was printed on an FDM machine with an additional Palette attachment that allowed for splicing filament colors together mid-print, resulting in a striped multicolor pattern that highlights the levels of the surface.

Acknowledgments

We thank the referees for their helpful comments, which improved the quality of this paper.

Thank you to Eric Stauffer, James Madison University Libraries, for processing and printing the filament-spliced model in Figure 19, and to Patrick Bishop, George Mason University, who printed the colorful models in Figures 1, 3, 7, and 9. Thanks also to Edmund Harriss, University of Arkansas, who took beautiful photographs of some of our models.

This material is based upon work supported by the National Science Foundation under Grant No. DMS-1439786 and the Simons Foundation Institute Grant Award ID 507536 while Evelyn Sander and Laura Taalman were in residence at the Institute for Computational and Experimental Research in Mathematics in Providence, Rhode Island, during the fall 2019 semester. Evelyn Sander's work was supported by a grant from the Simons Foundation/SFARI (636383, ES).

References

[1] K. Alligood, T. Sauer, and J. Yorke. *Chaos—An Introduction to Dynamical System*. Springer-Verlag, New York, 1996.

[2] V. S. Anishchenko and V. V. Astakhov. Experimental study of the mechanism of the appearance and the structure of a strange attractor in an oscillator with inertial nonlinearity. *Radiotekhnika i Elektronika*, 28:1109–1115, 1983. In Russian.

[3] V. S. Anishchenko and G. Strelkova. Irregular attractors. *Discrete Dynamics in Nature and Society*, 2(1):53–72, 1998.

[4] A. Arneodo, P.H. Coullet, E.A. Spiegel, and C. Tresser. Asymptotic chaos. *Phys. D,* 14(3):327–347, 1985.

[5] Chaotic Atmospheres. Math:rules, strange attractors. https://chaoticatmospheres.com /mathrules-strange-attractors, accessed February 2020.

[6] Guanrong Chen and Tetsushi Ueta. Yet another chaotic attractor. *Internat. J. Bifur. Chaos Appl. Sci. Engrg.,* 9(7):1465–1466, 1999.

[7] B.V. Chirikov. A universal instability of many-dimensional oscillator systems. *Phys. Rep.,* 52(5):263–379, 1979.

[8] Michael S. Gagliardo. 3D printing chaos. *Bridges 2018 Conference Proceedings,* pages 491–494, 2018. http://archive.bridgesmathart.org/2018/bridges2018-491.pdf.

[9] Peter Gray and Stephen K. Scott. *Chemical Oscillations and Instabilities,* International Series of Monographs on Chemistry, vol. 21. Oxford University Press, Oxford, U.K., 1990.

[10] M. Hénon and C. Heiles. The applicability of the third integral of motion: Some numerical experiments. *Astron. J.,* 69:73–79, 1964.

[11] Marius Kintel. *OpenSCAD: The Programmers Solid 3D CAD Modeller.*

[12] W.F. Langford. Numerical studies of torus bifurcations. In T. Küpper, H.D. Mittelmann, H. Weber, eds., *Numerical Methods for Bifurcation Problems.* International Series of Numerical Mathematics, vol. 70, pages 285–295. Birkhäuser, Basel, Switzerland, 1984.

[13] E.N Lorenz. Deterministic nonperiodic flow. *J. Atmos. Sci.,* 20:130141, 1963.

[14] Stephen K. Lucas, Evelyn Sander, and Laura Taalman. Supplementary materials for modeling dynamical systems for 3D printing. 2020. http://math.gmu.edu/~sander /EvelynSite/supplementary-materials-for.html.

[15] M.C. Mackey and L. Glass. Oscillation and chaos in physiological control systems. *Science,* 197(4300):287–289, 1977.

[16] James D. Meiss. *Differential dynamical systems,* Mathematical Modeling and Computation, vol. 22. Society for Industrial and Applied Mathematics (SIAM), Philadelphia, revised ed., 2017.

[17] Hinke M. Osinga and Bernd Krauskopf. Crocheting the Lorenz manifold. *Math. Intelligencer,* 26(4):25–37, 2004.

[18] O.E. Rossler. An equation for continuous chaos. *Phys. Lett. A,* 57:397–398, 1976.

[19] A. M. Rucklidge. Chaos in models of double convection. *Journal of Fluid Mechanics,* 237:209–229, 1992.

[20] David Ruelle. What is . . . a strange attractor? *Notices Amer. Math. Soc.,* 53(7):764–765, 2006.

[21] Evelyn Sander. Spinodal decomposition. *GMU Math Makerlab Blog,* 2015. http://gmumath maker.blogspot.com/2015/01/spinodal-decomposition.html.

Scientists Uncover the Universal Geometry of Geology

JOSHUA SOKOL

On a mild autumn day in 2016, the Hungarian mathematician Gábor Domokos arrived on the geophysicist Douglas Jerolmack's doorstep in Philadelphia. Domokos carried with him his suitcases, a bad cold, and a burning secret.

The two men walked across a gravel lot behind the house, where Jerolmack's wife ran a taco cart. Their feet crunched over crushed limestone. Domokos pointed down.

"How many facets do each of these gravel pieces have?" he said. Then he grinned. "What if I told you that the number was always somewhere around six?" Then he asked a bigger question, one that he hoped would worm its way into his colleague's brain. What if the world is made of cubes?

At first, Jerolmack objected. Houses can be built out of bricks, but Earth is made of rocks. Obviously, rocks vary. Mica flakes into sheets; crystals crack on sharply defined axes. But from mathematics alone, Domokos argued, any rocks that broke randomly would crack into shapes that have, on average, six faces and eight vertices. Considered together, they would all be shadowy approximations converging on a sort of ideal cube. Domokos had proved it mathematically, he said. Now he needed Jerolmack's help to show that this is what nature does.

"It was geometry with an exact prediction that was borne out in the natural world, with essentially no physics involved," said Jerolmack, a professor at the University of Pennsylvania. "How in the hell does nature let this happen?"

Over the next few years, the pair chased their geometric vision from microscopic fragments to rock outcrops to planetary surfaces and even to Plato's *Timaeus*, suffusing the project with an additional air of mysticism. The foundational Greek philosopher, writing around 360 BCE, had matched his five Platonic solids with five supposed elements: Earth, air, fire, water, and star stuff. With either foresight or luck, or a little of both, Plato paired cubes, the most stackable shape, with Earth. "I was like, oh, okay, now we're getting a little bit metaphysical," Jerolmack said.

But they kept finding cuboid averages in nature, plus a few non-cubes that could be explained with the same theories. They ended up with a new mathematical framework: a descriptive language to express how all things fall apart. When their paper was published earlier this year, it came titled like a particularly esoteric Harry Potter novel: "Plato's Cube and the Natural Geometry of Fragmentation."

Several geophysicists contacted by *Quanta* say the same mathematical framework might also help with problems like understanding erosion from cracked cliff faces, or preventing hazardous rock slides. "That is really, really exciting," said the University of Edinburgh geomorphologist Mikaël Attal, one of two scientists who reviewed the paper before publication. The other reviewer, the Vanderbilt geophysicist David Furbish, said, "A paper like this makes me think: Can I somehow make use of these ideas?"

All Possible Breaks

Long before he came to Philadelphia, Domokos had more innocuous mathematical questions.

Suppose you fracture something into many pieces. You now have a mosaic: a collection of shapes that could tile back together with no overlaps or gaps, like the floor of an ancient Roman bath. Further suppose that those shapes are all convex, with no indentations.

First Domokos wanted to see if geometry alone could predict what shapes, on average, would make up that kind of mosaic. Then he wanted to be able to describe all other possible collections of shapes you could find.

In two dimensions, you can try this out without smashing anything. Take a sheet of paper. Make a random slice that divides the page into two pieces. Then make another random slice through each of those two polygons. Repeat this random process a few more times. Then count up and average the number of vertices on all the bits of paper.

The Moeraki Boulders in New Zealand, a potential 3D Voronoi mosaic.
Daniel Lienert

For a geometry student, predicting the answer is not too hard. "I bet you a box of beer that I can make you derive that formula within two hours," Domokos said. The pieces should average four vertices and four sides, averaging to a rectangle.

You could also consider the same problem in three dimensions. About 50 years ago, the Russian nuclear physicist, dissident, and Nobel Peace Prize winner Andrei Dmitrievich Sakharov posed the same problem while chopping heads of cabbage with his wife. How many vertices should the cabbage pieces have, on average? Sakharov passed the problem on to the legendary Soviet mathematician Vladimir Igorevich Arnold and a student. But their efforts to solve it were incomplete and have largely been forgotten.

Unaware of this work, Domokos wrote a proof which pointed to cubes as the answer. He wanted to double-check, though, and he suspected that if an answer to the same problem already existed, it would be locked in an inscrutable volume by the German mathematicians Wolfgang Weil and Rolf Schneider, an 80-year-old titan in the field of geometry. Domokos is a professional mathematician, but even he found the text daunting.

"I found someone who was willing to read that part of the book for me and translate it back into human language," Domokos said. He found the theorem for any number of dimensions. That confirmed that cubes were indeed the 3D answer.

Now Domokos had the average shapes produced by splitting a flat surface or a three-dimensional block. But then a larger quest emerged. Domokos realized that he could also develop a mathematical description not just of averages, but of potentiality: Which collections of shapes are even mathematically possible when something falls apart?

Remember, the shapes produced after something falls apart are a mosaic. They fit together with no overlap or gaps. Those cut-up rectangles, for example, can easily tile together to fill in a mosaic in two dimensions. So can hexagons, in an idealized case of what mathematicians would call a Voronoi pattern. But pentagons? Octagons? They do not tile.

In order to properly classify mosaics, Domokos started describing them with two numbers. The first is the average number of vertices per cell. The second is the average number of different cells sharing each vertex. So in a mosaic of hexagonal bath tiles, for example, each cell is a hexagon, which has six vertices. And each vertex is shared by three hexagons.

The Geometry of Mars

To analyze a surface — in this case, the honeycombed surface of a Martian crater — researchers map out all edges and vertices. They count the number of vertices per cell, and how many cells share a given vertex.

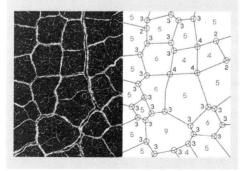

Samuel Velasco/*Quanta Magazine*. Based on graphics from doi.org/10.1073/pnas.2001037117; Martian surface: NASA/JPL-Caltech/University of Arizona. See also color insert.

In a mosaic, only certain combinations of these two parameters work, forming a narrow swath of shapes that could possibly result from something falling apart.

Once again, this full swath was fairly easy to find in two dimensions, but much harder in three. Cubes stack together well in 3D, of course, but so do other combinations of shapes, including those that form a 3D version of the Voronoi pattern. To keep the problem feasible, Domokos restricted himself to just mosaics with orderly, convex cells that share the same vertices. Eventually, he and the mathematician Zsolt Lángi devised a new conjecture that sketched out the curve of all possible three-dimensional mosaics like this. They published it in *Experimental Mathematics*, and "then I sent the whole thing to Rolf Schneider, who is of course the god," Domokos said.

"I asked him whether he wanted me to explain how I got this conjecture, but he reassured me that he knew," Domokos said, laughing.

The Cubic Cosmos

In three dimensions, most rocks break into cubes, with eight vertices per cell. The proposed map of allowable convex mosaics with orderly cells that meet at the same vertices forms only a narrow band.

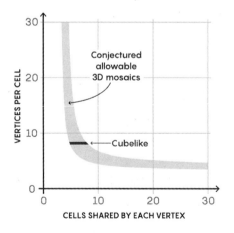

Samuel Velasco/*Quanta Magazine*. Based on graphics from doi.org/10.1073
/pnas.2001037117

"That meant like a hundred times more than being accepted in any journal."

More importantly, Domokos now had a framework. Mathematics offered a way to classify all the patterns that surfaces and blocks could break into. Geometry also predicted that if you fragmented a flat surface randomly, it would break into rough rectangles, and if you did the same in three dimensions, it would produce rough cubes.

But for any of this to matter to anyone other than a few mathematicians, Domokos had to prove that these same rules manifest themselves in the real world.

From Geometry to Geology

By the time Domokos swung through Philadelphia in 2016, he had already made some progress on the real-world problem. He and his colleagues at the Budapest University of Technology and Economics had gathered shards of dolomite eroded from a cliff face on the Hármashatárhegy mountain in Budapest. Over several days, a lab tech with no presuppositions about a universal conspiracy toward cubes painstakingly counted faces and vertices on hundreds of grains. On average? Six faces, eight vertices. Working with János Török, a specialist in computer simulations, and Ferenc Kun, an expert on fragmentation physics, Domokos found that cuboid averages showed up in rock types like gypsum and limestone as well.

With the math and the early physical evidence, Domokos pitched his idea to a stunned Jerolmack. "Somehow he's cast a spell, and everything else disappears for a moment," Jerolmack said.

Their alliance was a familiar one. Years ago, Domokos had won renown by proving the existence of the Gömböc, a curious three-dimensional shape that swivels into an upright resting position no matter how you push it. To see if Gömböcs existed in the natural world, he had recruited Jerolmack, who helped apply the concept to explain the rounding of pebbles on Earth and Mars. Now Domokos was again asking for help in translating lofty mathematical concepts into literal stone.

The two men settled on a new plan. To prove Plato's cubes actually appear in nature, they needed to show more than just a coincidental echo between geometry and a few handfuls of rock. They needed to consider all rocks and then sketch out a convincing theory of how

The Gömböc is a convex three-dimensional shape of uniform density that has a single stable equilibrium point. Domokos

abstract math could percolate down through messy geophysics and into even messier reality.

At first, "everything seemed to work," Jerolmack said. Domokos' mathematics had predicted that rock shards should average out to cubes. An increasing number of actual rock shards seemed happy to comply. But Jerolmack soon realized that proving the theory would require confronting rule-breaking cases, too.

After all, the same geometry offered a vocabulary to describe the many other mosaic patterns that could exist in both two and three dimensions. Off the top of his head, Jerolmack could picture a few real-world fractured rocks that did not look like rectangles or cubes at all but could still be classified into this larger space.

Perhaps these examples would sink the cube-world theory entirely. More promisingly, perhaps they would arise only in distinct circumstances and carry separate lessons for geologists. "I said I know that it doesn't work everywhere, and I need to know why," Jerolmack said.

Over the next few years, working on both sides of the Atlantic, Jerolmack and the rest of the team started plotting where real examples of broken rocks fell within Domokos' framework. When the team investigated surface systems that are essentially twodimensional—cracking permafrost in Alaska, a dolomite outcrop, and the exposed cracks of a granite block—they found polygons averaging four sides and four vertices, just like the sliced-up sheet of paper. Each of these geological cases seemed to appear where rocks had simply fractured. Here Domokos' predictions held up.

Another type of fractured slab, meanwhile, proved to be what Jerolmack had hoped for: an exception with its own distinct story to tell. Mud flats that dry, crack, get wet, heal, and then crack again have cells

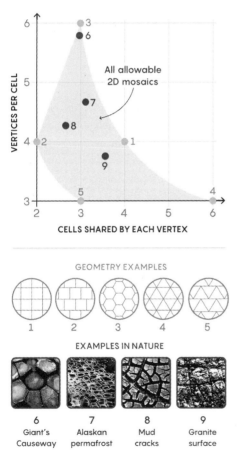

The Entire Tile Universe

What convex 2D shapes can fill a space without any overlaps or gaps? The possibilities can be mapped by counting the average number of vertices per shape and the number of shapes that share each vertex.

All allowable 2D mosaics

VERTICES PER CELL

CELLS SHARED BY EACH VERTEX

GEOMETRY EXAMPLES

1   2   3   4   5

EXAMPLES IN NATURE

| 6 | 7 | 8 | 9 |
| Giant's Causeway | Alaskan permafrost | Mud cracks | Granite surface |

Samuel Velasco/*Quanta Magazine*. Based on graphics from doi.org/10.1073 /pnas.2001037117; spot images: Lindy Buckley; Matthew L. Druckenmiller; Hannes Grobe; Courtesy of János Török. See also color insert.

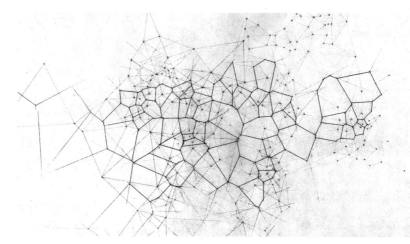

A Voronoi diagram separates a plane into individual regions, or cells, so that each cell consists of all points closest to a starting "seed" point. Fred Scharmen. See also color insert.

averaging six sides and six vertices, following the roughly hexagonal Voronoi pattern. Rock made from cooling lava, which solidifies downward from the surface, can take on a similar appearance.

Tellingly, these systems tended to form under a different type of stress—when forces pulled outward on a rock instead of pushing it in. The geometry revealed the geology. And Jerolmack and Domokos thought this Voronoi pattern, even if it was relatively rare, might also occur on scales far larger than they had previously considered.

Counting the Crust

Midway through the project, the team met in Budapest and spent three whirlwind days sprinting to incorporate more natural examples. Soon Jerolmack pulled up a new pattern on his computer: the mosaic of how Earth's tectonic plates fit together. Plates are confined to the *lithosphere*, a nearly two-dimensional skin on the surface of the planet. The pattern looked familiar, and Jerolmack called the others over. "We were like, oh wow," he said.

By eye, the plates looked as if they hewed to the Voronoi pattern, not the rectangular one. Then the team counted. In a perfect Voronoi mosaic of hexagons in a flat plane, each cell would have six vertices. The actual tectonic plates averaged 5.77 vertices. For a geophysicist, that was close enough to celebrate. For a mathematician, not so much. "Doug was getting into a good mood. He was working like hell," Domokos said. "I was getting in a depressed mood for the next day, because I was just thinking about the gap."

Domokos went home for the night, the difference still gnawing at him. He wrote down the numbers again. And then it hit him. A mosaic of hexagons can tile a plane. But Earth is not a flat plane. Think of a soccer ball, covered in both hexagons and pentagons. Domokos crunched the numbers for the surface of a sphere and found that on a globe, Voronoi mosaic cells should average 5.77 vertices.

This insight might help researchers answer a major open question in geophysics: How did Earth's tectonic plates form? One idea holds that plates are just a byproduct of burbling convection cells deep in the mantle. But an opposing camp holds that Earth's crust is a separate system—one that expanded, grew brittle, and cracked open. The observed Voronoi pattern of plates, reminiscent of much smaller mud flats, might support the second argument, Jerolmack said. "That's also what made me realize how important that paper was," Attal said. "It's really phenomenal."

A Revealing Break

In three dimensions, meanwhile, exceptions to the cuboid rule were rare enough. And they too could be produced by simulating unusual, outward-pulling forces. One distinctively non-cubic rock formation lies on the coast of Northern Ireland, where waves lap against tens of thousands of basalt columns. In Irish this is Clochán na bhFomhórach, the stepping-stones of a race of supernatural beings; the English name is the Giant's Causeway.

Crucially, those columns and other similar volcanic rock formations are six-sided. But Török's simulations produced Giant's Causeway-like mosaics as three-dimensional structures that had simply grown up from a two-dimensional Voronoi base, itself produced when volcanic rock cooled.

Zooming out, the team argues, you could classify most real fractured-rock mosaics using just Platonic rectangles, 2D Voronoi patterns,

The Giant's Causeway in Northern Ireland. Tyler Donaghy

and then—overwhelmingly—Platonic cubes in three dimensions. Each of these patterns could tell a geological story. And yes, with the appropriate caveats, you really could say the world is made of cubes.

"They did their due diligence in vetting their modeled forms against reality," said Martha-Cary Eppes, an earth scientist at the University of North Carolina, Charlotte. "My initial skepticism was allayed."

"The math is telling us that when we begin to fracture rocks, however we do it, whether we do it randomly or deterministically, there is only a certain set of possibilities," said Furbish. "How clever is that?"

Specifically, perhaps you could take a real fractured field site, count up things like vertices and faces, and then be able to infer something about the geological circumstances responsible.

"We have places where we have data we can think about in this way," said Roman DiBiase, a geomorphologist at Pennsylvania State University. "That would be a really cool outcome, if you can discern things that are more subtle than the Giant's Causeway, and hitting a rock with a hammer and seeing what the shards look like."

As for Jerolmack, after first feeling uncomfortable over a possibly coincidental connection to Plato, he has come to embrace it. After all, the Greek philosopher proposed that ideal geometric forms are central to understanding the universe but always out of sight, visible only as distorted shadows.

"This is literally the most direct example we can think of. The statistical average of all these observations is the cube," Jerolmack said. "But the cube never exists."

Bouncing Balls and Quantum Computing

DON MONROE

The history of science and mathematics includes many examples of surprising parallels between seemingly unrelated fields. Sometimes these similarities drive both fields forward in profound ways, although often they are just amusing.

In December, Adam Brown, a physicist at Google, described a surprisingly precise relationship between a foundational quantum-computing algorithm and a whimsical method of calculating the irrational number π. "It's just a curiosity at the moment," but "the aspiration might be that if you find new ways to think about things, that people will use that to later make connections that they'd not previously been able to make," Brown said. "It's very useful to have more than one way to think about a given phenomenon."

In a preprint posted online (but not yet peer-reviewed at press time), Brown showed a mathematical correspondence between two seemingly unconnected problems. One is the well-known Grover search algorithm proposed for quantum computers, which should be faster than any classical equivalent. The other is a surprising procedure in which counting the number of collisions between idealized billiard balls produces an arbitrarily precise value for π.

Quantum Algorithms

Quantum computing exploits quantum bits, or *qubits*, such as ions or superconducting circuits, that can simultaneously represent two distinct states. In principle, a modest number of qubits can represent and manipulate an exponentially larger number of combinations. Exploiting this possibility for computing seemed like a pipe dream, however, until researchers devised algorithms to extract useful information from the qubits. The first such algorithm, described in 1994 by Peter Shor, then at Bell Labs in New Jersey, efficiently finds the prime factors of a number, potentially cracking important cryptography schemes. The trick is to frame the problem as determining the repetition period of a sequence, essentially a Fourier transform, which can be found using global operations on an entire set of qubits.

The second fundamental algorithm, devised in 1996 by Lov Grover working independently at Bell Labs, operates quite differently. "Shor and Grover are the two most canonical quantum algorithms," according to Scott Aaronson of the University of Texas at Austin. "Even today, the vast majority of quantum algorithms that we know are recognizably either 'Shor-inspired' or 'Grover-inspired', or both."

Grover's algorithm manipulates the entire set of qubits simultaneously while preserving the relationships between them.

Grover's algorithm is often described as a database search, examining a list of N items to find the item that has a desired property. If the list is ordered by some label (for example, alphabetized), any label can be found by repeatedly dividing the list in successive halves, eventually requiring $\log_2 N$ queries. For an unsorted list, however, checking each item in turn requires, on average $N/2$ steps (and possibly as many as N).

Like other quantum algorithms, Grover's manipulates the entire set of qubits simultaneously while preserving the relationships between them (prematurely querying any qubit to determine its state turns it into an ordinary bit, squandering any quantum advantage). However, Grover showed that the desired item can generally be found with only $\frac{\pi}{4}\sqrt{N}$ global operations.

This improvement is less than that seen in Shor-style algorithms, which typically are exponentially faster than their classical counterparts. The Grover approach, however, can be applied to more general, unstructured problems, Brown notes.

The calculation starts with an equal admixture of all N qubits. The algorithm then repeatedly subjects all the qubits to two alternating manipulations. The first operation embodies the target: it inverts the state of a specific, but unknown, bit. The task is to determine which bit is altered, but not by measuring them all. The second operation does not require any information about the target. Grover found that each time this sequence is repeated, the weight of the target in the admixture increases (although this cannot be measured). After the correct number of repetitions, there is an extremely high chance a measurement will yield the correct result.

Bouncing Billiards

These sophisticated quantum manipulations may seem to have little relationship to bouncing billiard balls. Yet Brown, while working on issues related to Grover's algorithm, came across an animation by math popularizer Grant Sanderson that made him notice the similarities. In his paper, Brown shows there is a precise mapping between the two problems.

Sanderson's animation illustrates a surprising observation described in 2003 by Gregory Galperin, a mathematician at Eastern Illinois University in Charleston. In the paper "Playing Pool with π," he imagined two billiard balls moving without friction along a horizontal surface, bouncing off each other and off a wall on the left side in completely elastic collisions (which preserve their combined kinetic energy).

If the right-hand ball is sent leftward toward a second stationary ball that is much lighter, the smaller ball will be sent back toward the left-hand wall without slowing the larger ball much. The small ball will bounce off the wall, and then collide with the large one again, repeating this multiple times. Eventually the collisions will turn the large ball around until it finally escapes to the right faster than the small ball can pursue it.

The number of collisions needed before this escape can occur grows larger with the ratio of the mass of the large ball compared to the small one. If the masses are equal, it will take three bounces: the first transfers all motion from the right ball to the left one, which bounces off the wall and then transfers its momentum back to the right ball again. If the large ball is 100 times as massive, the process will take 31 bounces. If the mass ratio is 10,000, there will be 314 bounces. In a spectacularly impractical computation, for every increase of a factor of 100 in the mass ratio, the number of collisions (divided by the square root of the mass ratio) includes another digit to the digital representation of π, 3.1415926535 ...

Brown fortuitously encountered Sanderson's animation (which uses blocks instead of balls) when Grover's algorithm was fresh in his mind, and he recognized significant similarities between the two situations.

The two quantum operations, for example, correspond respectively to collisions between the balls and between the lighter ball and the wall. The mass ratio corresponds to the size of the database. Moreover, the final result was that the number of operations (or bounces) is proportional to π and to the square root of this size or mass ratio. (There are also two factors of two that reflect simple bookkeeping differences between the problems.)

Beyond the surprising connection between such different systems, what on earth is the number π doing in both cases? This irrational number is of course best known as the ratio of the circumference of a circle to its diameter, although it also appears in the proportions of ellipses, as well as higher dimensional objects, such as spheres. One way to define a circle is through an algebraic constraint on the horizontal and vertical coordinates x and y: The points of a circle with radius r are constrained to satisfy $x^2 + y^2 = r^2$.

As it turns out, both the billiard problem and the Grover algorithm have constraints of this form. Collisions of the balls or manipulations of the quantum system correspond to rotations along the circle defined by these constraints.

For example, for two billiard balls of mass m (with velocity v_m) and M (with velocity v_M), an elastic collision preserves their total kinetic energy, $\frac{1}{2} m v_m^2 + \frac{1}{2} M v_M^2$. Completely reversing the velocity of the larger ball requires a total "rotation" by $180°$ (π radians) in the plane with coordinates v_m and v_M.

Similarly, for quantum systems, the probability of observing a particular outcome is proportional to the square of the "wave function" corresponding to that outcome. The sum of the probability (squared amplitude) for the target and all other outcomes must be one.

Historical Examples of Connections

There is still the question, "Is this profound insight into the nature of reality, or is it just a sort of curiosity?" Brown said. "Maybe Grover search is telling us something profound about the nature of reality, and maybe the bouncing-ball thing is more of a curiosity, and maybe connecting them is more in the spirit of the second one than the first one."

Still, there have been numerous cases in physics, and especially in mathematics, where such connections have contributed profoundly to progress. For example, physicists have spent more than two decades exploring a surprising correspondence between strongly interacting multiparticle quantum systems and gravitational models incorporating curved space-time with one higher dimension. There is even hope the wormholes in space-time can help resolve paradoxes associated with quantum-mechanical "entanglement" of distant particles.

Mathematics has frequently advanced through connections between disparate fields. For example, Fermat's "last theorem," involving integer solutions of a simple equation, was only proved centuries later using methods from "elliptic curves." In another example, in January, computer scientists proved a theorem relating entanglement to Alan Turing's notion of decidable computations, which continues to shake up other seemingly unrelated fields.

For his part, Aaronson suspects the Grover-billiard correspondence, although "striking in its precision," is probably "just a cute metaphor (in the sense that I don't know how to use it to deduce anything about Grover's algorithm that we didn't already know). And that's fine."

Further Reading

Galperin, G., "Playing Pool with π: The Number π from a Billiard Point of View," *Regular and Chaotic Dynamics 8*, p. 375 (2003).

Brown, A.R., "Playing Pool with $|\psi\rangle$: From Bouncing Billiards to Quantum Search," *arXiv .org:1912.02207* (2019).

Sanderson, G., "How Pi Connects Colliding Blocks to a Quantum Search Algorithm," *Quanta* (2020).

Landmark Computer Science Proof Cascades through Physics and Math

KEVIN HARTNETT

In 1935, Albert Einstein, working with Boris Podolsky and Nathan Rosen, grappled with a possibility revealed by the new laws of quantum physics: that two particles could be entangled, or correlated, even across vast distances.

The very next year, Alan Turing formulated the first general theory of computing and proved that there exists a problem that computers will never be able to solve.

These two ideas revolutionized their respective disciplines. They also seemed to have nothing to do with each other. But now a landmark proof has combined them while solving a raft of open problems in computer science, physics, and mathematics.

The new proof establishes that quantum computers that calculate with entangled quantum bits or *qubits*, rather than classical 1s and 0s, can theoretically be used to verify answers to an incredibly vast set of problems. The correspondence between entanglement and computing came as a jolt to many researchers.

"It was a complete surprise," said Miguel Navascués, who studies quantum physics at the Institute for Quantum Optics and Quantum Information in Vienna.

The proof's co-authors set out to determine the limits of an approach to verifying answers to computational problems. That approach involves entanglement. By finding that limit, the researchers ended up settling two other questions almost as a byproduct: Tsirelson's problem in physics, about how to mathematically model entanglement, and a related problem in pure mathematics called the Connes embedding conjecture.

In the end, the results cascaded like dominoes.

"The ideas all came from the same time. It's neat that they come back together again in this dramatic way," said Henry Yuen of the University of Toronto and an author of the proof, along with Zhengfeng Ji of the University of Technology Sydney, Anand Natarajan and Thomas Vidick of the California Institute of Technology, and John Wright of the University of Texas, Austin. The five researchers are all computer scientists.

Undecidable Problems

Turing defined a basic framework for thinking about computation before computers really existed. In nearly the same breath, he showed that there was a certain problem computers were provably incapable of addressing. It has to do with whether a program ever stops.

Typically, computer programs receive inputs and produce outputs. But sometimes they get stuck in infinite loops and spin their wheels forever. When that happens at home, there is only one thing left to do.

"You have to manually kill the program. Just cut it off," Yuen said.

Turing proved that there is no all-purpose algorithm that can determine whether a computer program will halt or run forever. You have to run the program to find out.

"You've waited a million years and a program hasn't halted. Do you just need to wait 2 million years? There's no way of telling," said William Slofstra, a mathematician at the University of Waterloo.

In technical terms, Turing proved that this halting problem is undecidable—even the most powerful computer imaginable could not solve it.

After Turing, computer scientists began to classify other problems by their difficulty. Harder problems require more computational resources to solve—more running time, more memory. This is the study of computational complexity.

Ultimately, every problem presents two big questions: "How hard is it to solve?" and "How hard is it to verify that an answer is correct?"

Interrogate to Verify

When problems are relatively simple, you can check the answer yourself. But when they get more complicated, even checking an answer can be an overwhelming task. However, in 1985 computer scientists realized it is possible to develop confidence that an answer is correct even when you cannot confirm it yourself.

The method follows the logic of a police interrogation. If a suspect tells an elaborate story, maybe you cannot go out into the world to confirm every detail. But by asking the right questions, you can catch your suspect in a lie or develop confidence that the story checks out.

In computer science terms, the two parties in an interrogation are a powerful computer that proposes a solution to a problem—known as the *prover*—and a less powerful computer that wants to ask the prover questions to determine whether the answer is correct. This second computer is called the *verifier*.

To take a simple example, imagine you are color-blind, and someone else—the prover—claims two marbles are different colors. You cannot check this claim by yourself, but through clever interrogation, you can still determine whether it is true.

Put the two marbles behind your back and mix them up. Then ask the prover to tell you which is which. If they really are different colors, the prover should answer the question correctly every time. If the marbles are actually the same color—meaning they look identical—the prover will guess wrong half the time.

"If I see you succeed a lot more than half the time, I'm pretty sure they're not" the same color, Vidick said.

By asking a prover questions, you can verify solutions to a wider class of problems than you can on your own.

In 1988, computer scientists considered what happens when two provers propose solutions to the same problem. After all, if you have two suspects to interrogate, it is even easier to solve a crime, or verify a solution, since you can play them against each other.

"It gives more leverage to the verifier. You interrogate, ask related questions, cross-check the answers," Vidick said. If the suspects are telling the truth, their responses should align most of the time. If they are lying, the answers will conflict more often.

Similarly, researchers showed that by interrogating two provers separately about their answers, you can quickly verify solutions to an even larger class of problems than you can when you only have one prover to interrogate.

Computational complexity may seem entirely theoretical, but it is also closely connected to the real world. The resources that computers need to solve and verify problems—time and memory—are fundamentally physical. For this reason, new discoveries in physics can change computational complexity.

"If you choose a different set of physics, like quantum rather than classical, you get a different complexity theory out of it," Natarajan said.

The new proof is the end result of twenty-first-century computer scientists confronting one of the strangest ideas of twentieth-century physics: entanglement.

The Connes Embedding Conjecture

When two particles are entangled, they do not actually affect each other—they have no causal relationship. Einstein and his co-authors elaborated on this idea in their 1935 paper. Afterward, physicists and mathematicians tried to come up with a mathematical way of describing what entanglement really meant. Yet the effort came out a little muddled. Scientists came up with two different mathematical models for entanglement—and it was not clear that they were equivalent to each other.

In a roundabout way, this potential dissonance ended up producing an important problem in pure mathematics called the Connes embedding conjecture. Eventually, it also served as a fissure that the five computer scientists took advantage of in their new proof.

The first way of modeling entanglement was to think of the particles as spatially isolated from each other. One is on Earth, say, and the other is on Mars; the distance between them is what prevents causality. This is called the *tensor product model*.

But in some situations, it is not entirely obvious when two things are causally separate from each other. So mathematicians came up with a second, more general way of describing causal independence.

When the order in which you perform two operations does not affect the outcome, the operations "commute": 3×2 is the same as 2×3. In this second model, particles are entangled when their properties are correlated but the order in which you perform your measurements does not matter: Measure particle A to predict the momentum of particle B or vice versa. Either way, you get the same answer. This is called the *commuting operator model* of entanglement.

Both descriptions of entanglement use arrays of numbers organized into rows and columns called matrices. The tensor product model uses matrices with a finite number of rows and columns. The commuting operator model uses a more general object that functions like a matrix with an infinite number of rows and columns.

Over time, mathematicians began to study these matrices as objects of interest in their own right, completely apart from any connection to the physical world. As part of this work, a mathematician named Alain Connes conjectured in 1976 that it should be possible to approximate many infinite-dimensional matrices with finite-dimensional ones. This is one implication of the Connes embedding conjecture.

The following decade, a physicist named Boris Tsirelson posed a version of the problem that grounded it in physics once more. Tsirelson conjectured that the tensor product and commuting operator models of entanglement were roughly equivalent. This makes sense, since they are theoretically two different ways of describing the same physical phenomenon. Subsequent work showed that because of the connection between matrices and the physical models that use them, the Connes embedding conjecture and Tsirelson's problem imply each other: Solve one, and you solve the other.

Yet the solution to both problems ended up coming from a third place altogether.

Game Show Physics

In the 1960s, a physicist named John Bell came up with a test for determining whether entanglement was a real physical phenomenon, rather than just a theoretical notion. The test involved a kind of game whose outcome reveals whether something more than ordinary, non-quantum physics is at work.

Computer scientists would later realize that this test about entanglement could also be used as a tool for verifying answers to very complicated problems.

But first, to see how the games work, let us imagine two players, Alice and Bob, and a 3-by-3 grid. A referee assigns Alice a *row* and tells her to enter a 0 or a 1 in each box so that the digits sum to an odd number. Bob gets a *column* and has to fill it out so that it sums to an even

Quantum Games

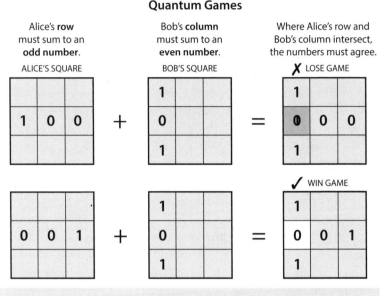

| Alice's **row** must sum to an **odd number.** | Bob's **column** must sum to an **even number.** | Where Alice's row and Bob's column intersect, the numbers must agree. |

FIGURE 1. An example of the "Magic Square" game, which players can win more frequently than expected with the aid of entangled particles. Courtesy of Lucy Reading-Ikkanda/*Quanta Magazine*.

number. They win if they put the same number in the one place her row and his column overlap. They are not allowed to communicate.

Under normal circumstances, the best they can do is win 89% of the time. But under quantum circumstances, they can do better.

Imagine Alice and Bob split a pair of entangled particles. They perform measurements on their respective particles and use the results to dictate whether to write 1 or 0 in each box. Because the particles are entangled, the results of their measurements are going to be correlated, which means their answers will correlate as well—meaning they can win the game 100% of the time.

So if you see two players winning the game at unexpectedly high rates, you can conclude that they are using something other than classical physics to their advantage. Such Bell-type experiments are now called "nonlocal" games, in reference to the separation between the players. Physicists actually perform them in laboratories.

"People have run experiments over the years that really show this spooky thing is real," said Yuen.

As when analyzing any game, you might want to know how often players can win a nonlocal game, provided they play the best they can. For example, with solitaire, you can calculate how often someone playing perfectly is likely to win.

But in 2016, William Slofstra proved that there is no general algorithm for calculating the exact maximum winning probability for all nonlocal games. So researchers wondered: Could you at least approximate the maximum winning percentage?

Computer scientists have homed in on an answer using the two models describing entanglement. An algorithm that uses the tensor product model establishes a floor, or minimum value, on the approximate maximum winning probability for all nonlocal games. Another algorithm, which uses the commuting operator model, establishes a ceiling.

These algorithms produce more precise answers the longer they run. If Tsirelson's prediction is true, and the two models really are equivalent, the floor and the ceiling should keep pinching closer together, narrowing in on a single value for the approximate maximum winning percentage.

But if Tsirelson's prediction is false, and the two models are not equivalent, "the ceiling and the floor will forever stay separated," Yuen said. There will be no way to calculate even an approximate winning percentage for nonlocal games.

In their new work, the five researchers used this question—about whether the ceiling and floor converge and Tsirelson's problem is true or false—to solve a separate question about when it is possible to verify the answer to a computational problem.

Entangled Assistance

In the early 2000s, computer scientists began to wonder: How does it change the range of problems you can verify if you interrogate two provers that share entangled particles?

Most assumed that entanglement worked against verification. After all, two suspects would have an easier time telling a consistent lie if they had some means of coordinating their answers.

But over the past few years, computer scientists have realized that the opposite is true: By interrogating provers that share entangled particles, you can verify a much larger class of problems than you can without entanglement.

"Entanglement is a way to generate correlations that you think might help them lie or cheat," Vidick said. "But in fact you can use that to your advantage."

To understand how, you first need to grasp the almost otherworldly scale of the problems whose solutions you could verify through this interactive procedure.

Imagine a graph—a collection of dots (vertices) connected by lines (edges). You might want to know whether it is possible to color the vertices using three colors, so that no vertices connected by an edge have the same color. If you can, the graph is "three-colorable."

If you hand a pair of entangled provers a very large graph, and they report back that it can be three-colored, you will wonder: Is there a way to verify their answer?

For very big graphs, it would be impossible to check the work directly. So instead, you could ask each prover to tell you the color of one of two connected vertices. If they each report a different color, and they keep doing so every time you ask, you will gain confidence that the three-coloring really works.

But even this interrogation strategy fails as graphs grow really big—with more edges and vertices than there are atoms in the universe. Even the task of stating a specific question ("Tell me the color of XYZ vertex") is more than you, the verifier, can manage: The amount of data required to name a specific vertex is more than you can hold in your working memory.

But entanglement makes it possible for the provers to come up with the questions themselves.

"The verifier doesn't have to compute the questions. The verifier forces the provers to compute the questions for them," Wright said.

The verifier wants the provers to report the colors of connected vertices. If the vertices are not connected, then the answers to the questions will not say anything about whether the graph is three-colored. In other words, the verifier wants the provers to ask correlated questions: One prover asks about vertex ABC and the other asks about vertex XYZ. The hope is that the two vertices are connected to each other, even though neither prover knows which vertex the other is thinking about. (Just as Alice and Bob hope to fill in the same number in the same square even though neither knows which row or column the other has been asked about.)

If two provers were coming up with these questions completely on their own, there would be no way to force them to select connected, or correlated, vertices in a way that would allow the verifier to validate their answers. But such correlation is exactly what entanglement enables.

"We're going to use entanglement to offload almost everything onto the provers. We make them select questions by themselves," Vidick said.

At the end of this procedure, the provers each report a color. The verifier checks whether they are the same or not. If the graph really is three-colorable, the provers should never report the same color.

"If there is a three-coloring, the provers will be able to convince you there is one," Yuen said.

As it turns out, this verification procedure is another example of a nonlocal game. The provers "win" if they convince you their solution is correct.

In 2012, Vidick and Tsuyoshi Ito proved that it is possible to play a wide variety of nonlocal games with entangled provers to verify answers to at least the same number of problems you can verify by interrogating two classical computers. That is, using entangled provers does not work against verification. And last year, Natarajan and Wright proved that interacting with entangled provers actually expands the class of problems that can be verified.

But computer scientists did not know the full range of problems that can be verified in this way. Until now.

A Cascade of Consequences

In their new paper, the five computer scientists prove that interrogating entangled provers makes it possible to verify answers to unsolvable problems, including the halting problem.

"The verification capability of this type of model is really mind-boggling," Yuen said.

But the halting problem cannot be solved. And that fact is the spark that sets the final proof in motion.

Imagine you hand a program to a pair of entangled provers. You ask them to tell you whether it will halt. You are prepared to verify their answer through a kind of nonlocal game: The provers generate questions and "win" based on the coordination between their answers.

If the program does in fact halt, the provers should be able to win this game 100% of the time—similar to how if a graph is actually three-colorable, entangled provers should never report the same color for two connected vertices. If it does not halt, the provers should only win by chance—50% of the time.

That means if someone asks you to determine the approximate maximum winning probability for a specific instance of this nonlocal game, you will first need to solve the halting problem. And solving the halting problem is impossible. Which means that calculating the approximate maximum winning probability for nonlocal games is undecidable, just like the halting problem.

This in turn means that the answer to Tsirelson's problem is no—the two models of entanglement are not equivalent. Because if they were, you could pinch the floor and the ceiling together to calculate an approximate maximum winning probability.

"There cannot be such an algorithm, so the two [models] must be different," said David Pérez-García of the Complutense University of Madrid.

The new paper proves that the class of problems that can be verified through interactions with entangled quantum provers, a class called MIP*, is exactly equal to the class of problems that are no harder than the halting problem, a class called RE. The title of the paper states it succinctly: "MIP* = RE."

In the course of proving that the two complexity classes are equal, the computer scientists proved that Tsirelson's problem is false, which, due to previous work, meant that the Connes embedding conjecture is also false.

For researchers in these fields, it was stunning that answers to such big problems would fall out from a seemingly unrelated proof in computer science.

"If I see a paper that says MIP* = RE, I don't think it has anything to do with my work," said Navascués, who co-authored previous work tying Tsirelson's problem and the Connes embedding conjecture together. "For me it was a complete surprise."

Quantum physicists and mathematicians are just beginning to digest the proof. Prior to the new work, mathematicians had wondered whether they could get away with approximating infinite-dimensional matrices by using large finite-dimensional ones instead. Now, because the Connes embedding conjecture is false, they know they cannot. "Their result implies that's impossible," said Slofstra.

The computer scientists themselves did not aim to answer the Connes embedding conjecture, and as a result, they are not in the best position to explain the implications of one of the problems they ended up solving.

"Personally, I'm not a mathematician. I don't understand the original formulation of the Connes embedding conjecture well," said Natarajan.

He and his co-authors anticipate that mathematicians will translate this new result into the language of their own field. In a blog post announcing the proof, Vidick wrote, "I don't doubt that eventually complexity theory will not be needed to obtain the purely mathematical consequences."

Yet as other researchers run with the proof, the line of inquiry that prompted it is coming to a halt. For more than three decades, computer scientists have been trying to figure out just how far interactive verification will take them. They are now confronted with the answer, in the form of a long paper with a simple title and echoes of Turing.

"There's this long sequence of works just wondering how powerful" a verification procedure with two entangled quantum provers can be, Natarajan said. "Now we know how powerful it is. That story is at an end."

Dark Data

DAVID J. HAND

Many readers of *Significance* will be familiar with the statistical story behind the 1986 *Challenger* Space Shuttle disaster. This involved just seven data points—and seven deaths. The story hinges around a graph showing the relationship between air temperature at seven previous shuttle launches and whether the seals on the joints between segments of the rocket boosters were distressed. There appeared to be no relationship beyond random variability. What was missing from the graph, however, were points for all those previous launches which had had *no* problems with the seals. Including these extra points in the graph led to a completely different picture—and inference: higher air temperatures at launch were associated with fewer problems.[1] I think that nobody, looking at the complete data set, would have decided that a launch should go ahead with the forecast temperature for the next day. The mistaken inference of no relationship led to the booster rocket exploding at nine miles of altitude, killing all seven astronauts.

The data missing from the *Challenger* plot are a very simple example of dark data. And, as that example illustrates—and as I describe in my book, *Dark Data: Why What You Don't Know Matters*[2]—the consequences can be catastrophic. They can involve lost fortunes, damaged reputations, and even death.

The Unknown

Dark data are data you do not have. They might be data you thought you had, or hoped to have, or wished you had. But they are data you don't have. You might be aware that you do not have them, or you might be unaware. In any case, as a result of these dark data, the data missing from your view of the world, you are at risk of misunderstanding, of drawing incorrect conclusions about the way the world works, of making poor predictions, of getting things wrong, just as in the *Challenger* example.

An example of a domain where dark data are less obvious but frequent is medical diagnosis. Diagnostic definitions and thresholds change over time, as understanding grows. This can reveal previously concealed depths of illness. For example, according to Huang et al., worldwide about a third of cases of diabetes are undiagnosed, and Huang attributes this to shortcomings of the traditional plasma glucose measurement procedures.[3] Likewise, autism, originally introduced into the *Diagnostic and Statistical Manual of Mental Disorders* in 1980, had its diagnostic definition changed in 1987 and 1994 in such a way that more people fell within its scope. And, of course, the problem of changing definitions leading to cases appearing or disappearing is not unique to medicine. Think of changing definitions of unemployment, for example. At the time of writing, the world is trying to cope with the Covid-19 pandemic, illustrating many kinds of dark data, but in particular missing counts of people who have the disease without (perhaps yet) showing symptoms.

Statisticians are very familiar with certain kinds of dark data. Non-response in surveys is a classic example, where it is not uncommon to find that people who refuse to take part are those who would have answered in a particular way. This is seen in pre-election polls, when social pressures might disincline people to admit to certain positions. Non-response is a worldwide problem, but is nicely illustrated by the UK Labour Force Survey, where overall response rates have declined from 55.5% to 38.6% over the past 10 years (bit.ly/2QJEFfy). Tools for attempting to tackle survey non-response date back many decades, but while statisticians can do amazing things, they cannot perform miracles. At some point, the dark data cast a serious shadow over any conclusions one might hope to draw.

Hidden in Plain Sight

While dark data in surveys are a familiar problem, the uncomfortable fact is that dark data creep in everywhere.

Dark data in the form of missing or unmeasured characteristics can be particularly pernicious. Think of the difficulty of testing discrimination if gender data are not available.

Concealed changes of data collection methods can also lead to making previously visible data invisible, or the reverse. A search engine optimization company, Moz, maintains a web page of all the confirmed (and unconfirmed) changes made to Google's search algorithm over the years (bit.ly/2UBRxp1). According to Moz, "In 2018, [Google] reported an incredible 3,234 updates—an average of almost 9 per day, and more than 8 times the number of updates in 2009. While most of these changes are minor, Google occasionally rolls out a major algorithmic update . . . that affects search results in significant ways."

Survivor bias is another relatively familiar dark data phenomenon in some domains. For example, straightforward rankings of investment fund performance will include only those funds which survived throughout the assessment period, the others constituting invisible dark data. Since, in general, it will be the worst performing funds that drop out, overall measures of performance will be biased upward unless steps are taken to allow for this. And the phenomenon can be significant. In a study by Vanguard, an investment management company, only just over half of funds survived the 15-year study period (https://studylib.net/doc/18317319/the-mutual-fund-graveyard--an-analysis-of-dead-funds). Readers may be familiar with similar issues involving dropouts in clinical trials.

Even if you are correct in supposing that there are no selection distortions in your data, dark data can obscure the truth in other ways. Short of simple counting, no measurements are *completely* accurate—to an infinite number of decimal places. That means that your observations are necessarily approximate, with the values you are analyzing rounded to some extent. And the darkening of data by rounding and heaping of numbers to particular values can lead to mistaken conclusions, by biasing overall summary statistics or through mistaken classifications if a threshold is used (see the box "How Rounding or Heaping Can Lead to Problems").

A phenomenon related to rounding is truncation, where it is known only that a true value is greater than or less than some threshold. A mercury thermometer will not record a value less than the freezing point of mercury, for example, and a bathroom weighing scale will not record a value greater than its upper limit.

My book gives a taxonomy of 15 types of dark data, those described above and others, including missing entire variables and distortions due

---

HOW ROUNDING OR HEAPING CAN LEAD TO PROBLEMS

The tendency to round numbers to convenient approximate values, ending in 5 or 10 for example, can obviously lead to distortions in estimates of means, variances, and skewness. But even the normal rounding implicit in recording data to a finite number of decimal places can lead to problems. The failure of the American Patriot Missile battery in Dharan in Saudi Arabia to intercept an Iraqi Scud missile which killed 28 and injured 100 was attributed to the rounding to 24 bits of the unending binary expansion of 1/10 (of a second) when this error was aggregated over the 100 hours the system had been operating.

---

to changes over time. Unfortunately, different types of dark data are not mutually exclusive: identifying one dark data problem in your observations does not mean that others (dark dark data, I suppose) are not also there. The different kinds can work together in a diabolical synergy. I argue that the data you don't know can be at least as important as the data you do know, at least if you hope to draw valid conclusions.

Lighting the Way

I wrote the book primarily to raise awareness of the dangers of dark data. All too often, analyses are undertaken without sufficient thought given to the genesis and provenance of the data. Machine learning algorithms, for example, always give an output, regardless of how partial or misleading the input data are, and eagerness to discover the conclusion discourages people from losing time by rigorously thinking about the possible shortcomings of the data.

But the outlook is not all gloomy, and tools have been developed for tackling dark data.

The first step is to detect the dark data—or really, to detect the hole revealing the absence of data. Sometimes this is easy: missing fields in a survey response are a simple gap (and an example of what Donald

---

THE PROBLEM WITH SUBSTITUTING AVERAGES FOR MISSING VALUES

Substituting the average observed age for people who don't give their age in a survey assumes there is no systematic difference between the missing and observed ages. It is also likely to yield an underestimate of the variance of age in the population and distort relationships of age with other characteristics.

Rumsfeld called "known unknowns"). In other cases, it is tougher: missing responses to a blog question are not so apparent—we don't know who might have responded (Rumsfeld's "unknown unknowns").

Then, once you have recognized that there could be something crucial missing, the key strategy for tackling dark data is to *use what you do know about what you don't know.*

Many simple methods for doing this have been proposed, and even integrated into statistical software packages. They include things like using only the complete cases, or using all the values recorded on individual variables, or substituting the average of the observed values. Unfortunately, while such methods can occasionally be satisfactory, in most cases they are not: simple and obvious is likely to lead one into deeper shadows as far as dark data are concerned (see "The Problem with Substituting Averages for Missing Values").

Effective methods rely on understanding—or assuming something about—the nature of the mechanism which led to the dark data, and more elaborate methods are based on modeling the relationships between observed values and dark data, leading to tools such as multiple imputation and the expectation-maximization algorithm. But as I said earlier, statisticians cannot perform miracles. At some point, we must make assumptions about why the data are dark.

The Cover of Darkness

So far, I have described dark data that have arisen accidentally—or at least not by the intention of the researchers. But sometimes data are darkened deliberately. Fraudsters do this all the time, of course. But so do you: you use dark data (your passwords) to protect data from prying eyes. And that represents a positive use of dark data.

Much more sophisticated positive uses of dark data also occur in what I call *the strategic application of ignorance*. We hide which treatment each group of patients receive from patients and researchers in a clinical trial. We use randomized response methods to extract sensitive information. We generate data *which might have been* but which were not when we carry out simulations. We create imaginary copies of misclassified cases when we use boosting in classification algorithms. We create possible new data sets when we slightly perturb data to regularize it to yield more robust models. We conjure up possible past data when we write down a Bayesian prior. And so on.

Returning to my opening story, the *Challenger* disaster, a statistician looking at the original graph of seven data points should have been suspicious. The graph appears to show that *every* previous launch had at least one seal with a problem: five had a single seal problem, one had two seal problems, and one had three. There were *none* with no problems. That is intrinsically surprising—if the seal problems are independent, then by chance alone, we would expect the occasional launch with no problems. That suspicion would have led to a search, revealing that all previous launches with no problems had been omitted from the graph. It would have revealed dark data.

References

1. Dalal, S. (2016) From risk to resiliency. *Significance*, **13**(1), 42–43.
2. Hand, D. J. (2020) *Dark Data: Why What You Don't Know Matters.* Princeton, NJ: Princeton University Press.
3. Huang, H., Peng, G., Lin, M., Zhang, K., Wang, Y., Yang, Y., Zuo, Z., Chen, R., and Wang, J. (2013) The diagnostic threshold of HbA1c and impact of its use on diabetes prevalence—A population-based survey of 6898 Han participants from southern China. *Preventive Medicine*, **57**, 345–350.

Analysis in an Imperfect World

MICHAEL WALLACE

What did you eat yesterday? This is a simple question, with a seemingly straightforward answer. In my case, I had toast for breakfast, a sandwich for lunch, and pasta for dinner. Do snacks count? I had an apple during the afternoon and, if you must know, a doughnut with my morning coffee (I was teaching first thing). Oh, do you need to know what I drank as well? I think along with the coffee there were two—no, maybe three—cups of tea? I drank water throughout the day too, but didn't measure it ...

Whenever we conduct statistical analyses, we make numerous assumptions. Many of these assumptions are familiar to anyone who has taken an introductory statistics class. We may assume our data "follow a particular distribution" (such as the famous "bell curve") or that the effect of a treatment is the same for different patients. Arguably the most common assumption of all, however, is barely mentioned: that the measurements we take are without error. In other words, when we ask for someone's height, or weight, or blood pressure, or simply what they had for dinner, we assume the measurement we get is exactly the same as whatever we are *trying* to measure.

In some cases, this is a reasonable assumption. Age, at least where reliable birth records are available, is accurate, as are a handful of other variables, such as biological sex or employment status. We may also be prepared to assume that our measurements are "close enough" to what we are trying to observe that it is not worthwhile worrying about error. However, such cases tend to be the exception rather than the rule, and to assume our measurements are perfect (or sufficiently "near perfect"), when in reality they are not, can have grave statistical consequences.

A dramatic example occurred in the study of cervical cancer and human papillomavirus (HPV) infection.[1] The link between these two

HPV AND CANCER

In their 1994 paper, Schiffman and Schatzkin studied the relationship between HPV infection and cervical intraepithelial neoplasia (CIN, a precursor to cervical cancer).[1] To do so, they analyzed two case–control studies, where individuals with CIN (the cases) were compared to those without (the controls). The first study used an older method to measure HPV infection, a version of the Southern blot, while the second used a newer method called polymerase chain reaction (PCR). Although some good tests were available at the time of the first study—including a more reliable version of the Southern blot—they were highly labor-intensive and thus unsuitable for large-scale epidemiological studies.

Repeat testing of HPV infection was available on a subset of individuals in each study, whereby evidence of the fallibility of the Southern blot was derived. Of 51 participants successfully retested in the Southern blot study, there were 11 "discordant pairs": individuals who received differing results. In contrast, all 43 of those retested in the PCR study produced the same diagnosis.

For both studies, the risk of developing CIN was assessed using odds ratios. An *odds ratio* is a numeric measure of how much one's risk of an outcome increases in the presence of another factor, compared to if that factor is absent. In this case, odds ratios were used to compare the risk of developing CIN between those with and without HPV infection. Here, the difference between the Southern blot and PCR measures of HPV infection was stark: the odds ratio for CIN associated with a positive HPV test was 3.7 using Southern blot, but 20.1 using PCR. Using the more reliable test revealed a much stronger association between HPV infection and CIN.

Later work went on to establish a causative relationship between certain strains of HPV and CIN, leading to the development of a vaccination against HPV infection. Today, the Centers for Disease Control and Prevention in the United States and the United Kingdom National Health Service, along with many other countries' health agencies, recommend vaccination against HPV.

conditions is now well established, with HPV vaccinations a common recommendation to mitigate this risk. For many years, however, this relationship was dismissed, thanks in part to a number of studies that suggested no clear association between the two conditions. The cause? Measurement error.

An Inconvenient Truth

Measurement error occurs when the value of an observation does not equal the "truth" we, the analysts, are trying to observe. In some cases, this is easy to understand: If you are 160 cm tall, that represents the truth, but if I make a muddle of my tape measure, I may end up with a measurement of 161 cm. Often, though, the mere definition of "truth" is surprisingly difficult to pin down. Your weight fluctuates throughout the day, week, month, and year. If I want to use weight as a variable in an analysis, what is the "true" weight I am trying to measure? Your weight right now, the average across the day, or something else altogether?

Furthermore, such errors come in many forms. At its simplest, we may view what we observe as equal to the truth plus some "random" noise. This is the *classical* measurement error model and includes the appealing scenario where we are just as likely to under- as to overestimate what we wish to observe. Often, however, errors may be systematic in nature. A common example is *white coat hypertension*, where blood pressure readings—often made in doctors' offices or other high-stress environments—are typically elevated. Asking people how often they smoke cigarettes or drink alcohol (or eat doughnuts) may result in other systematic errors.

A different type of model we may be willing to assume for the error is known as *Berkson error*. For example, if we wish to study the effect of air pollution on the health of a population, we might attempt to measure the levels of particulate matter in the environment. Air quality monitors record such data, but it is likely unknown to how much particulate matter an individual person might be exposed. We can therefore view the "truth" (how much particulate matter an individual breathes in, for example) as equal to the error-prone observation from the air quality monitor, but with noise added. This is in contrast to the classical sense, where the error-prone observation may be thought of as the truth with noise added. What type of error we encounter, whether it is systematic or truly random, and what variables in our analysis it affects, all impact both the problems it can cause and the methods available to us to address them.

While Berkson and classical errors are often perceived to dichotomize the measurement error universe, there are more nuanced considerations. Errors in the outcome, such as misclassification of disease diagnosis, can pose unique challenges. Moreover, the outcome itself can sometimes influence the accuracy of reporting, creating what is known as *differential measurement error*. For example, someone who has received a lung cancer diagnosis may then overestimate their historical cigarette consumption as a result, potentially leading to exaggerated effect estimates.

In the aforementioned case of HPV and cervical cancer, measurement error occurred in the detection of HPV infection itself. Identifying a patient as HPV-free when the patient was not, or vice versa, constitutes measurement error: the value we observe (whether HPV infection is detected or not) is different from the truth (whether the patient is actually infected). Sometimes, such error is referred to as *misclassification* because a categorical variable is under consideration.

Regardless of what we call it, measurement error represents a ubiquitous problem, especially in medical statistics, where much of the literature relies on measuring attributes of people. Beyond health research, it can be just as problematic, affecting laboratory observations in chemistry or physics, or questions of econometrics (consider, for example, asking people about their salaries, or the value of their houses). Regardless of discipline, the word "error" can sometimes cause consternation among researchers, especially those using the best available tools or techniques. Unfortunately, truly perfect measurements may be impossible to take. Despite its prevalence, however, measurement error remains a relatively new—and comparatively specialist—area of research. Moreover, it is often ignored in practice.

But why?

A Convenient Untruth

For years, measurement error has been ignored because of the claim that its impact is "not that bad." But this general belief stems from a convenient result that only holds in a few special cases. To explain, a common goal in research is to estimate the strength of a relationship between two (or more) variables, such as between HPV infection and cancer. When errors are consistent with the classical measurement error model, and not systematic (in other words, when what we observe is equal to the truth plus some noise that, on average, is zero), we can construct analyses where measurement error at worst leads us to *underestimate* the strength of such relationships (*see* "Attenuation and a Post Hoc Fix"). One perception is that the overestimation of an effect is a far greater crime than underestimation: Imagine, for example, if a pharmacist over- rather than underestimated the efficacy of a new drug. There are also some statistical tests that can remain theoretically valid despite such attenuation (that is, reduction in magnitude) of estimated effects. Consequently, we may also underestimate the negative impact of measurement error itself.

Whether these observations alone are sufficient grounds to view measurement error as an ignorable concern is a matter for debate in itself. Indeed, attenuated effects are not necessarily harmless. A particularly notable example comes from a 1990 article in *The Lancet*.[2] The authors studied the association between blood pressure and the risks of stroke and heart disease. Measurement error, and the aforementioned attenuation of effects, led to these risks being severely underestimated. Moreover, regardless of whether attenuation is ever acceptable, it can be shown—and especially in the case of systematic error—that the strength of an effect can be *overestimated* as well. Anticipating the extent to which measurement error must be addressed can swiftly become a complex challenge.

The tendency to overlook the impact of measurement error due to the belief that it only causes a weakening in effects is at best puzzling and at worst deeply concerning. In the case of HPV and cancer, for example, such a weakening of the relationship delayed implementation of health practices designed to prevent the former causing the latter (*see* "HPV and Cancer"). More generally, the idea that this could be the only effect of errors in the increasingly complex models used by statisticians and non-statisticians alike would be laughable were the consequences not potentially so severe.

That is not to say, however, that the underappreciation of measurement error stems only from this misapprehension. Another factor is

ATTENUATION AND A POST HOC FIX

If we consider simple linear regression, where a "straight-line" relationship is sought between a response Y and an exposure X, we are trying to find a relationship between the two of the form

$$Y = \alpha + \beta X + \epsilon$$

where ϵ represents "noise." Viewed this way, the strength of the relationship between X and Y is given by β: If it is large and positive, then Y will increase quickly as X increases, and vice versa. If X represented dosage of a drug, and Y the resulting health of a patient, we can see how estimating β can tell us how well the drug is working. Standard statistical methods allow us to estimate β and determine the relationship between X (the *covariate*) and Y (the *outcome*).

If our covariate is measured with error, then instead of X we might observe $X^* = X + U$, where U represents the measurement error (and is assumed to be independent of X). Larger values of U would mean our observed value X^* is further from X, suggesting more severe measurement error. If we carry out our statistical analysis, we use X^* in our equation, instead of X, which can affect our resulting estimate of β.

In this case, if we assume that U follows a normal distribution with mean zero and variance σ_u^2, and that X has variance σ_x^2, then our standard analysis would estimate not β but

$$\beta^* = \frac{\sigma_x^2}{\sigma_x^2 + \sigma_u^2} \beta$$

Because $\sigma_x^2 / (\sigma_x^2 + \sigma_u^2) < 1$, this means $|\beta^*| \le |\beta|$: Our estimate will be *attenuated*. In other words, we will underestimate the strength of the relationship between X and Y. This is illustrated by the true relationship being given by a steeper slope in Figure 1.

In this setting, however, if we can estimate σ_x^2 and σ_u^2, we can easily estimate β: Carry out a "naive" analysis to obtain an estimate of β^*, then multiply it by our estimate of $(\sigma_x^2 + \sigma_u^2)/\sigma_x^2$. This correction (a form of regression calibration) "undoes" the

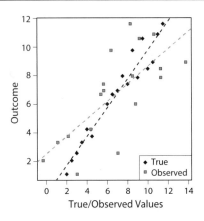

FIGURE 1. The attenuation effect: Measurement error in the observed data can reduce the size of the estimated effect, as indicated by the slope (β) of the fitted lines. See also color insert.

attenuating effect of measurement error in this situation, returning us to an estimate of the true relationship between X and Y.

The slope of a straight-line fit to a data set can give an indication of the strength of the relationship between the two variables: A steeper line suggests a stronger relationship. Here in Figure 1, a straight-line fit to the error-prone observations (squares) is less steep than that fit to the true values (diamonds), and the measurement error would therefore lead to an attenuated effect estimate. Intuitively, we can view the points as being "stretched" horizontally (the error either increases or decreases the x-axis values) but not vertically, as the outcome is assumed to be measured without error. This leads to a shallower fitted line.

The curious reader may wonder what would happen if, in contrast, the outcome were subject to measurement error of the same form detailed here, while the covariate remained error-free. In such a scenario, it can be shown that the strength of the relationship would be maintained, but with greater uncertainty in its accuracy. Again, however, such a convenient result cannot be relied upon in more general settings.

the perception that it is a problem too difficult to solve. When I raise the issue of measurement error with collaborators on scientific projects, for example, I am sometimes met with a resigned shrug that the measurements they have are the best they can do (or sometimes offense at the mere suggestion that their measurements are not perfect!). Furthermore, while statistical methodologies to analyze error-prone data do exist, they are seldom known to a research team. With proper planning, these techniques can be easily incorporated into an analysis, and ideally into the design of the study itself.

More Data

One of the biggest obstacles we face in the study of measurement error is in characterizing the size and structure of the measurement error itself. If a set of patients each has their blood pressure measured once, there is not much we can do to detect—or correct for—the measurement error that is surely present. In general, we need some additional data from which we can learn more.

Typically, the cheapest form of such data is known as *repeated measures* or *replicates*. As the names suggest, rather than taking a single measurement, we take repeated measures on at least a subset of our sample, giving us valuable insights into how accurate our data are. For example, if the difference between two measurements is very small, this would suggest our measurements are fairly accurate. A large discrepancy would give us greater concern that measurement error may be particularly problematic. At a very simplistic level, we could imagine taking two measurements and using their average, but more sophisticated techniques can take advantage of statistical theory to improve our analyses further.

A major problem repeated measures cannot (necessarily) solve is that of systematic error. After all, if your blood pressure is *always* elevated when visiting the doctor, an average of several measurements will also be elevated. The ideal source of additional information, therefore, are *validation* data, where some measurements are known to be perfect. This can sometimes be accomplished through more expensive or invasive procedures, or through the development of new techniques (as with the improvement in HPV testing; *see* "HPV and Cancer"). Often, however, this is simply not possible, because either no practical measurement mechanism exists, or whatever we are trying to measure in the first place is not well defined. In the latter scenario, we sometimes use what is known as a "gold standard," where researchers determine the best measure they can reasonably obtain given practical or other constraints. Nevertheless, in such cases we must be careful in our interpretation and ensure that any associations are clearly set in the context of our gold standard and not lazily associated with some (unobtainable) "perfect" measurement.

Besides replication and validation data, there are a handful of other ways to acquire information about the measurement error itself, but these tend to be less reliable or more complicated to implement. A common problem arises in practice where, owing to measurement error not being anticipated in the design of a study, no additional data are available to address this problem at the analysis stage. In such cases, our last resort is often *external* data: measurements—or knowledge of measurement error—from other sources or studies. Of course, with this comes additional uncertainty about how accurately such external information translates to our own setting, but it may be our only choice.

All Correct

There are therefore many options available to us, and what decisions we make at both the design and analysis stages of any study should be informed by measurement error considerations. We should begin by asking what types of error we anticipate, and in which variables, and then identify where in our design we can make useful accommodations.

For illustration, let us suppose that you are conducting a study and, having read this article and been convinced of the perils of measurement error, you decide to anticipate its effects. You determine that while one of your variables will be measured with error, it will be with random, additive noise (in other words, classical and not systematic). Luckily, while validation data are too expensive to obtain, repeated measures can be collected relatively cheaply, so you measure every subject twice.

In the simplest of settings (where the myth of attenuation is, in fact, a reality), we can carry out a typical, or "naive," analysis and apply a retrospective *method of moments* correction. For this, we use our additional data to estimate some aspects of our problem—including the size of our measurement error—and then use a simple formula to correct our naive results (*see* "Attenuation and a Post Hoc Fix").

In more general settings, a popular technique is known as *regression calibration*. For this, we again conduct our usual analysis, such as a linear or logistic regression, but replace the error-prone observations with our "best guess" of the truth for each individual (or, at least, our best guess based on what is observed or measurable). This is a little like using the mean of each subject's observed values, but it takes the additional information about the size (and distribution) of the measurement error into account to yield more accurate results. It can even be used in the case of systematic error, if we know enough about that aspect of our error process.

Another popular method, known as *simulation extrapolation* (SIMEX), takes a rather different but nevertheless intuitively appealing approach. Having learned about the structure of the measurement error in the system, we are able to simulate new data sets as if they were subject to more and more severe (that is, larger) measurement error. This can lead to a pattern in our resulting estimates: for instance, if the strength of a relationship becomes gradually smaller as the measurement error increases, we can follow this pattern backward to the hypothetical scenario where no error is present. SIMEX is usually applied in the case of classical error, but can be used in other settings, such as with spatial data (*see* "Simulation Extrapolation").

Naturally, a whole host of techniques are available for addressing the problem of measurement error in practice. If you have a data set where measurement error is present, chances are there is a correction method that will fit your particular scenario. Many of these methods have implementations in common software environments, such as Stata's *merror* package, *simex* in R, and SAS macros from the U.S. National Cancer Institute. Simpler approaches, such as the method of moments correction and regression calibration, can even be implemented directly with relative ease (especially if you have a friendly statistician at hand). Of course, the use of any statistical methods must be approached with care, and close attention must be paid to any underlying assumptions upon which they rely.

SIMULATION EXTRAPOLATION

Simulation extrapolation begins with a "naive" analysis (step 1) that does not take measurement error into account, producing a (likely biased) estimate of the effect of interest. The size of the measurement error is then estimated so that new data sets may be simulated where additional error is added to the already error-prone measurements (step 2). For example, a data set with an "error multiplier" of 2 is one where the observed error-prone measurements have been replaced with simulated values estimated to suffer from twice as much measurement error as the original data. For each of these new data sets, the effect is estimated. This pattern is then used to extrapolate back to the scenario where no error is present (step 3).

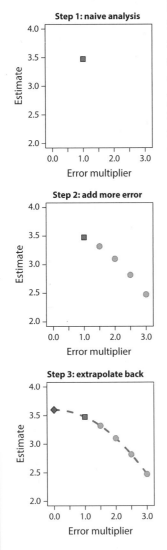

FIGURE 2. Simulation extrapolation: By adding more measurement error to the data, the relationship between error and effect estimate can be studied and extrapolated back to the case of zero error. See also color insert.

Future Perfect

Regardless of your statistical background (or the friendliness of any statistically inclined colleagues), the most important lesson is that accounting for measurement error is by no means impossible. Indeed, it is often the case that with careful planning, measurement error can be minimized or avoided altogether. This can be achieved, for example, through the use of more precise measurement techniques or by selecting variables that are less susceptible to mismeasurement.

Failing that, additional data—though not essential—vastly enhance our capacity to address the problems measurement error might cause. What is more, they can often be collected with comparatively little cost. From here, there is a vast—and expanding—array of methods available for correcting our analyses.

Ultimately, however, those who do not take advantage of such approaches must face a hard truth: Failure to take measurement error into account can completely invalidate their findings. While we can seldom expect perfection in measurement or modeling, there is much that can—and should—be done to get us a little bit closer to it.

Acknowledgments

This article was written on behalf of the measurement error and misclassification topic group of the international STRengthening Analytical Thinking for Observational Studies (STRATOS) Initiative (stratos-initiative.org). STRATOS aims to provide accessible and accurate guidance documents for relevant topics in the design and analysis of observational studies. The topic group is led by Laurence Freedman and Victor Kipnis, with members Hendriek Boshuizen, Raymond Carroll, Veronika Deffner, Kevin Dodd, Paul Gustafson, Ruth Keogh, Helmut Küchenhoff, Pamela Shaw, Anne Thiebaut, Janet Tooze, and Michael Wallace.

References

1. Schiffman, M. H., and Schatzkin, A. (1994) Test reliability is critically important to molecular epidemiology: An example from studies of human papillomavirus infection and cervical neoplasia. *Cancer Research*, **54**, 1944s–1947s.

2. MacMahon, S., Peto, R., Cutler, J., Collins, R., Sorlie, P., Neaton, J., Abbott, R., God-win, J., Dyer, A., and Stamle, J. (1990) Blood pressure, stroke, and coronary heart disease. Part 1, Prolonged differences in blood pressure: Prospective observational studies corrected for the regression dilution bias. *Lancet*, **335**, 765–774.

Further Reading

For more on measurement error, *see*:

- Carroll, R. J., Ruppert, D., Stefanski, L. A., and Crainiceanu, C. M. (2006) *Measurement Error in Nonlinear Models: A Modern Perspective*, 2nd ed. Boca Raton, FL: Chapman and Hall/CRC.

- Coggon, D., Rose, G., and Barker, D. J. P. (2003) Measurement error and bias. In *Epidemiology for the Uninitiated*, 5th ed. London: BMJ Publishing Group. bit.ly/mErrorBias.

- Gustafson, P. (2004) *Measurement Error and Misclassification in Statistics and Epidemiology: Impacts and Bayesian Adjustments*. Boca Raton, FL: Chapman and Hall/CRC.

A Headache-Causing Problem

J. H. CONWAY, M. S. PATERSON,
AND U. S. S. R. MOSCOW

Setting. When just N men have been gathered together in the room shown in Figure 1, the blind lady umpire makes the following (true) announcement:

"We are all, as we know, infinitely intelligent and honorable people. Now the janitor, acting on my instructions, has attached to your foreheads small discs bearing the usual notations for various non-negative integers, in such a way that each of you can see the number on everyone else's head, but not that on your own. The sum of all these numbers is one of the numbers you can now see me writing on the blackboard."

"I regret the slight discomfort this proceeding must have caused you—fortunately, the theorem of [1] assures us that it will only last for a few more moments. I will now question each of you in turn, and at the first 'Yes' answer we can all go out and enjoy what is left of this lovely afternoon."

She now asks:

"Arthur, can you deduce solely from this information what number is written on your disc?"

If Arthur's reply is "No", she will turn to the next man, and ask:

"Bertram, can you deduce from the above information together with Arthur's reply what number must be written on *your* disk?"

If Bertram in turn says "No", she will question Charles, Duncan, and so on, perhaps reaching the Nth man:

"Engelbert, can you now deduce your number from the given information together with all the replies you have now heard?"

If even Engelbert says "No", she will return to Arthur, and continue cyclically, always asking the same question:

Figure 1   The game (2,2,2 | 6,7,8)

FIGURE 1. The game (2, 2, 2 | 6, 7, 8).

"Are you now able to deduce your number solely from the given information and the replies you have heard so far?" until the game is terminated by a "Yes" reply.

Statement of the theorem. *If the number of numbers written on the blackboard is less than or equal to the number N of men, the game will terminate after a finite number of questions.*

The disproof (commenced). It has become customary (*see*, e.g., [1]) to present the disproof of this theorem before its proof. The disproof runs as follows.

To demolish this perfectly preposterous proposition, it will suffice of course to disprove any particular case. We shall take the case when $N = 3$, the number on every head is 2, and the numbers on the blackboard are 6, 7, and 8, and establish conclusively that it will never end. We shall refer to this as the case (2, 2, 2 | 6, 7, 8).

It will help us to imagine Charles at his breakfast table that morning.

"Oh dear, yet another invitation from Zoe. She's a lovely girl, and intelligent too, but I do wish she wouldn't keep dragooning us into playing those ridiculous party games.

"Wonder what it'll be this time? I'd better just run through the infinitely many possibilities so we'll be able to get it over with as quickly as possible. If it's Charades, I'll repeat exactly the one I did last time—they're *bound* to get that first go. If it's Hunt-the-Slipper, I'll ...

... ...

... But then again, she might just be thinking of playing the game described so wittily in [1]. In that case, she's as likely as not to use the particular case called (2, 2, 2 | 6, 7, 8) in the handy notation of that reference. What should my reactions be?"

Charles's Argument. "Let me think now. Since I see two heads numbered 2, I will know from the start that my number will be 2, 3, or 4. Let's consider these cases.

"*If my number's* 2, Arthur will have concluded that *his* number was 2, 3, or 4, and since each of these is consistent with all he was told, he'll have to say 'No'.

"Bertram's then in a similar position. He'll think 'If I have 2, Arthur, by Charles's argument of the previous paragraph, will say 'No'. If 3, Arthur will instead have been able to conclude only that his number was 1, 2, or 3, and to say 'No' because he'll only know that his number is 0, 1, or 2. Must remember that she said *non-negative* numbers so that 0 is allowed.'

"Since in this case Bertram won't be able to eliminate even one of his three possibilities, 2, 3, 4, he'll be forced to say 'No'. That disposes of the case when my number is 2.

"*If my number is* 3, Arthur fairly obviously still says 'No'. Bertram will know probably in condensed form:

'I can see $A = 2$, $C = 3$, so I know $B = 1, 2$, or 3.
If $B = 1$, A'll've been torn between 2, 3, 4, so'll've said 'No'.
If $B = 2$, A'll've been torn between 1, 2, 3, so'll've said 'No'.
If $B = 3$, he'll've been torn between 0, 1, 2, so'll still've said 'No'.
   I must therefore say 'No' myself, since all three cases are consistent with A's 'No' answer.'

"Bertram and Arthur will also both say 'No' when my number is 3. I think I can prove along the same lines that they'll both say 'No' even when it's 4. But I don't need to check this—my first answer must be

'No' because both 2 and 3 are consistent with the two 'No' answers I'm sure to hear.

"I plainly don't need to consider much more of this stuff—I reckon we'll all go home after saying 'No' half a dozen times, and I still won't know what my number is."

The disproof completed. Charles's argument, and various portions of it, can be used to complete with absolute rigor that each of the three players knows from the start that each of the first three answers is going to be "No". *So if they all know what those answers are going to be, what information can they possibly gain by hearing them ritually intoned?* At the start of the second round, they will have learned nothing that they did not already know, and so the game will obviously go on forever.

The proof (commenced). We might as well make it clear now that Zoe, the blind lady umpire, is herself ignorant of the numbers fixed to those heads, although she knows, of course, what numbers she wrote on the blackboard. In the interests of good order, she will naturally list all situations that are compatible with the numbers she has heard up to any given time and will strike a situation off her list when and only when she knows that the corresponding game would terminate at the current question. Of course, she knows just when this will be, for being infinitely intelligent she can perfectly well imagine herself in the position of the player she is currently addressing in any possible situation.

By a *possible situation* we mean of course an N-tuple of numbers

$$(a, b, c, \ldots , n)$$

which might be the numbers on the respective heads of

$$(A, B, C, \ldots , N)$$

and would have caused "No" answers to all questions before the current one. We shall call such a situation *ongoing* only if the answer to the current question will also be "No".

We claim that Zoe can work out exactly which situations are ongoing by the following argument:

"Let me suppose that the current question is addressed to B. Then certainly

i) I *cannot* strike off $(a, b, c, \ldots, n)$ if there's an *accompanying* situation $(a, b', c, \ldots, n)$ still present with the same values of $a, c, \ldots, n$ but with b' differing from b, for then B cannot eliminate either of the numbers b and b'.

ii) I *can* strike off $(a, b, c, \ldots, n)$ if there's no such accompanying situation, for then B, who can see the numbers $a, c, \ldots, n$ will know that b is the only possible value for his number.

"So when I receive a 'No' answer from B (say), I must strike off those and only those points $(a, b, c, \ldots, n)$ from my N-dimensional record that are unaccompanied by any other point

$$(a, b', c, \ldots, n)$$

in the B-direction."

Since Zoe's argument covers all cases, we can now follow her algorithm to discover exactly which situations will cause the game to end at any given time.

The case $(2, 2, 2 \mid 6, 7, 8)$. Before we resume the proof for the general case, we illuminate the fate of all games of the form $(a, b, c \mid 6, 7, 8)$ in Figure 2. This figure shows an orthogonal projection of the set of all points in (A, B, C) space that yield sums of 6, 7, or 8, together with Zoe's notes as to the number of questions whose "Yes" answer terminates the game.

The four entries "19," singled out in Figure 3, enable us to verify both of Charles's predictions about the game $(2, 2, 2 \mid 6, 7, 8)$.

The proof completed. We return now to the discussion of the general case. No matter what numbers are written on the blackboard, provided that there are at most N of them, the total number of initial situations will surely be finite. We shall show that no one of these situations can remain ongoing at every possible moment in time.

For otherwise, the set of permanently ongoing situations would be non-empty and have the property that every one of its points would be accompanied in every coordinate direction. It will suffice to show that any such set of points in N dimensions has at least $N + 1$ distinct sums, and we can verify this by induction.

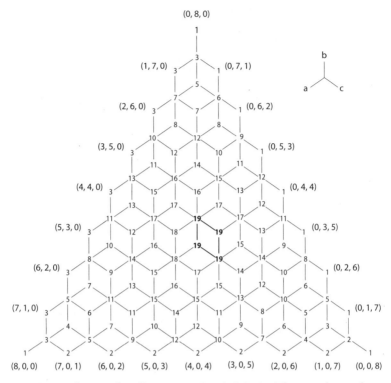

FIGURE 2. Zoe's notes for all games $(a, b, c \mid 6, 7, 8)$. The coordinate directions for (a, b, c) are indicated in the upper right.

FIGURE 3. At the end of the game (this detail is highlighted in the center of Figure 2).

Let a_0 be the least value of the A-coordinate of any permanently ongoing situation in an N-man game. Then the tuples of $N - 1$ numbers

$$(b, c, \ldots, n)$$

for which $(a_0, b, c, \ldots, n)$ is permanently ongoing in this game will themselves form a permanently ongoing set in an $N - 1$ man game, and so will have at least $(N - 1) + 1 = N$ distinct sums. Let

$$s_0 = a_0 + b_0 + c_0 + \ldots + n_0$$

be the greatest of these, arising from the permanently ongoing situation

$$(a_0, b_0, c_0, \ldots, n_0).$$

Then there is a permanently ongoing situation

$$(a, b_0, c_0, \ldots, n_0)$$

accompanying this in the A direction with $a \neq a_0$, and so $a > a_0$ since a_0 was minimal. The accompanying situation therefore has coordinate sum greater than any of those already found and establishes that there must be at least $N + 1$ distinct coordinate sums in all.

Application to a problem of Fermat. The problem referred to is Fermat's famous assertion that

$$a \geq 1, b \geq 1, c \geq 1, n \geq 3 \quad \Rightarrow \quad a^n \neq b^n + c^n \ (\star)$$

for rational integers a, b, c, n.

Now it is well known that for every proposition P, we have

$$(P \text{ and not-}P\,) \Rightarrow (0 = 1).$$

Taking P to be the proposition discussed so disarmingly in **[1]**, and applying *modus ponens*, we deduce that

$$0 = 1.$$

Now adding 1 to both sides of this, we obtain

$$1 = 2,$$

which we prefer to write in the more revealing form

$$1^3 = 1^3 + 1^3.$$

Thus the lexicographically first case of $(\star)$ is disproved. The authors cannot resist the remark that this would surely have been noticed earlier had not modern teaching methods preferred the elaboration of grandiose general theories to the inculcation of elementary arithmetical skills.

Acknowledgments

The work described here was carried out when the first and second named authors enjoyed the hospitality of the third. The second and third authors are indebted to the first for expository details. The first and third authors gratefully remark that without the constant stimulation and witty encouragement of the second author, this paper was completed.

Note

This article was originally presented in a "Festschrift" published on the occasion of Hendrik Lenstra's Ph.D. defense in Amsterdam on May 18, 1977. The privately published book was entitled *Een Pak met een Korte Broek: Papers presented to H. W. Lenstra, Jr. on the Occasion of the Publication of his "Eulidische Getallenlichamen."* It was edited by P. van Emde Boas, J. K. Lenstra, F. Oort, A. H. G. Rinnooy Kan, and T. J. Wansbeek, who have approved the publication of this article in this volume. We are excited to present this paper to you, and, as P. van Emde Boas stated, "This paper has had a major impact on the early developments of epistemic logic in Amsterdam."

References

[1] Conway, J. H., Paterson, M. S., and Moscow, U. S. S. R. (1977). A Headache-Causing Problem." In *The Best Writing on Mathematics 2021*, M. Pitici, Ed., Princeton, NJ, Princeton University Press.

A Zeroth Power Is Often a Logarithm Yearning to Be Free

SANJOY MAHAJAN

As a calculus student many decades ago, I found disturbing the $n = -1$ exception when integrating a power law,

$$\int x^n dx = \begin{cases} \dfrac{x^{n+1}}{n+1} + C & (n \neq -1) \\ \ln x + C & (n = 1) \end{cases} \tag{1}$$

I appreciated that mathematically the caveat saved us from dividing by zero. But I found it mysterious how, only at $n = -1$, the result, which had been a power law like the integrand, suddenly turned into a new kind of function, $\ln x$.

The mystery returned a few years later when, as physics students, we computed the electrostatic potential of simple sources: a point, a line, and a sheet of charge. Said another way, we computed the potential of a point source in three dimensions (the point charge), in two dimensions (the line charge beloved of electrical engineers for pole–zero plots), and in one dimension (the surface charge). The potential misbehaved in a familiar way. In three dimensions, it was r^{-1}. In one dimension, it was r^1. Based on these data, it should be r^{2-n} in n dimensions. But in two dimensions, for which this theory predicts r^0, we get $\ln r$.

In graduate school, as I began haltingly thinking like a physicist rather than a physics student, I wondered more about the whole idea of an exception limited to exactly a point. The physical world abhors a discontinuity. Pretending that a formula is valid everywhere except at one point ($n = -1$) is unphysical.

But perhaps the exception in Equation (1) was hinting at an approximation: that $\ln x$ could be approximated by a power law. With $n = -1 + \epsilon$ (for $\epsilon \approx 0$) and with integration range $[1, x]$, Equation (1) says

$$\ln x \approx \frac{x^\epsilon - 1}{\epsilon} \tag{2}$$

A high-order root (and a subtraction) approximates a logarithm! For example, with $\epsilon = 0.1$ and $x = 10$, the right-hand side of Equation (2) is approximately 2.6, only 12% larger than $\ln 10$. In general, the fractional error is roughly $0.5\epsilon \cdot \ln x$, so the approximation is reasonable when $x \ll e^{2/\epsilon}$.

The $\epsilon \to 0$ limit of Equation (2) reveals the zeroth power as a logarithm yearning to be free. Indeed, a logarithm grows more slowly than any positive power. Thus, if the logarithm is any power law at all, the exponent has to be the smallest positive number, often known as zero. The electrostatic potential in two dimensions (the potential of a line charge), by all rights a zeroth power, is such a logarithm.

Here are two further examples of this idea, one biophysical and one mathematical (with application to orbits).

The biophysical example is the conversion of light intensity into neural signals in the retina. The retina, perhaps our first back-of-the-envelope reasoner, hides from us irrelevant details about the world. As an illustration, take the front page of the *New York Times* outside on a sunny day, measure the light intensity above the giant black masthead letters and above the surrounding blank newsprint, and do the same indoors. The black letters outdoors give off far more light than the white areas indoors! Yet, to our eyes, the newspaper looks the same indoors and outdoors.[1]

This magical illumination invariance could be achieved by division: Measure the light intensity locally and divide it by a suitable local average of the light intensity. This ratio would be the same indoors and outdoors. Alas, in the real world of neural hardware with voltages and currents, division is a difficult operation—but subtraction is easy.

Enter the logarithm. If the neural signal can be made proportional to the light intensity's logarithm (after making the intensity suitably dimensionless), the retina is home free. Although a retinal receptor cannot ask a floating-point coprocessor to compute logarithms for it, it computes instead an approximate fourth root. This root is computed from two pieces. First, a highly cooperative process, requiring the simultaneous binding of several calcium ions, computes a fourth power (of calcium concentration). Second, using calcium feedback, this computation is wrapped in a negative-feedback loop to compute the fourth

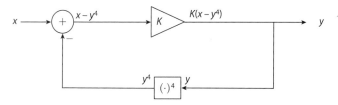

FIGURE 1. Producing a fourth root using negative feedback. Chasing the signal around the loop produces the constraint equation $K(x - y^4) = y$, where x is the input signal, y is the output signal, and K is the forward-path gain. As long as K is large, the solution is $y = x^{1/4}$ (a slight generalization of Black's formula).

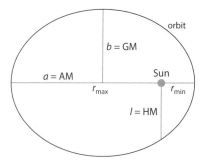

FIGURE 2. The various means in a Kepler orbit. The semimajor axis (a) is the arithmetic mean (AM) of the perihelion (r_{min}) and the aphelion (r_{max}). The semiminor axis (b) is their geometric mean (GM). The semi-latus rectum (l) is their harmonic mean (HM).

root (Figure 1)—which approximates a logarithm[2] in the same sense that the right-hand side of Equation (2) approximates ln x.

The second prodigal logarithm answers a question about the geometric mean (GM) raised in an earlier article.[3] There, the GM was one of several means in a Kepler orbit (Figure 2). The other means are all power means, where the kth power mean is defined as

$$M_k \equiv \left(\frac{a^k + b^k}{2} \right)^{1/k} \qquad (3)$$

The arithmetic mean (AM) is M_1, the harmonic mean (HM) is M_{-1}, and the root-mean-square (RMS) is M_2. But what power mean, if any, is the GM?

A hint comes from lining up several known power means from greatest to least,

$$\underbrace{\text{RMS}}_{M_2} \geq \underbrace{\text{AM}}_{M_1} \geq \underbrace{\text{GM}}_{M_?} \geq \underbrace{\text{HM}}_{M_{-1}} \qquad (4)$$

The GM lies between M_1 and M_{-1}, so perhaps it's the zeroth power mean.

For a proof, take the $k \to 0$ limit of Equation (3). The important ingredient is that, for tiny k,

$$z^k = e^{k \ln z} \approx 1 + k \ln z \tag{5}$$

Applying Equation (5) with $z = a$ and $z = b$ turns Equation (3) into

$$M_k \approx \left[\frac{(1 + k \ln a) + (1 + k \ln b)}{2} \right]^{1/k}$$
$$= \left(1 + k \frac{\ln a + \ln b}{2} \right)^{1/k} \tag{6}$$

Inverting Equation (5), now setting $\ln z = (\ln a + \ln b)/2$, turns the final expression in Equation (6) into

$$\exp \left(\frac{\ln a + \ln b}{2} \right) \tag{7}$$

which is $\sqrt{ab}$: the usual route to calculating the geometric mean.

Although I've sinned against Wheeler's first moral principle[4] by calculating before understanding the answer, I'll redeem myself by obeying a moral corollary:[5] After solving a problem, ask yourself, "What one sentence would have most helped my earlier self crack this problem?"

Here, a clue to the sentence is the appearance of logarithms in Equation (7), the alternative route to computing the geometric mean. When combined with the idea that the logarithm is a zeroth power and with its mirror image, that an exponential is an infinite power, Equation (7) becomes

$$\left(\frac{a^0 + b^0}{2} \right)^{1/0} \tag{8}$$

which is M_0, the zeroth power mean. Wherefore the one sentence: A logarithm is a zeroth power in disguise.

Notes

1. A separate question is whether, after many further stages of neural processing turning the ink patterns into sentences and their meaning, one should believe what's been printed with all that ink.

2. This simplified description is itself the result of many other approximations. More details and related calculations can be found in Sanjoy Mahajan, "Order of magnitude physics: A textbook with applications to the retinal rod and to the density of prime numbers," Ph.D. dissertation (California Institute of Technology, 1998), pp. 165–180.

3. Sanjoy Mahajan, "Don't demean the geometric mean," *Am. J. Phys.* **87**, 75–77 (2019), after Equation (4).

4. Edwin F. Taylor and John Archibald Wheeler, *Spacetime Physics*, 2nd ed. (W. H. Freeman, New York, 1992), p. 20.

5. Because of its close relation to Wheeler's first moral principle, I've often attributed this one-sentence question—a wonderful means (sorry) of deliberate practice—also to John Wheeler. If a gentle reader can confirm or refute this attribution, please let me or the editor know.

The Bicycle Paradox

Stan Wagon

Mark Levi's recent book review [3] in *The Mathematical Intelligencer* reports on an error made by V. I. Arnold. But the exact nature of the error needs clarification. Here is what Arnold wrote:

> **What Force Drives a Bicycle Forward?** The lower pedal of a bicycle standing still on a horizontal floor is pulled back. Which way does the bicycle go, and in what direction does the pulled back lower pedal move with respect to the floor?

Arnold's diagram is reproduced in Figure 1. Here x is the (infinitesimal) distance traveled left by the pedal relative to the bicycle. His conclusion is that for a particular choice of parameters, the bike (z) moves forward a distance $5x$ with respect to the ground and the pedal therefore moves forward $5x - x = 4x$.

Arnold's error is subtle. He is correct that if x moves left relative to the bike (i.e., the pedal rotates clockwise), then the bike moves forward by $5x$. But his error is in the assumption that the pedal rotates clockwise when pulled left. The opposite is true! This counterintuitive fact is known as the bicycle paradox. Readers unfamiliar with this surprising behavior are encouraged to find a bicycle and try it.

The book editor thought that Arnold was thinking of a cyclist sitting on the saddle in the standard position and pushing the pedals. If so, all that Arnold says is correct. But this is surely false (as Levi noted); the normal riding of a bike uses shoes and friction to propel one forward, and there is no mystery to resolve. Arnold simply was unaware of the bicycle paradox. His mathematics is correct, but he made a physics error (Figure 2).

In his review, Levi wrote that "Arnold's answer is correct in reference to the bike itself rather than the pedal." This comment is false for

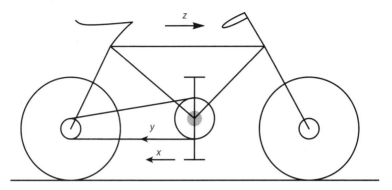

FIGURE 1. Arnold's diagram.

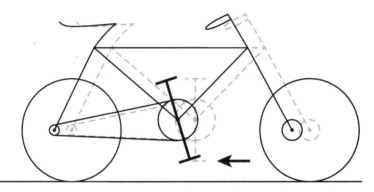

FIGURE 2. A clockwise push on the pedal (dashed lines) makes it go counter-clockwise, and the bicycle moves backward (solid black lines).

any standard bike, though it is correct if the bike has an exceptionally low gear ratio. Here we present the geometry and physics that provide a simple and complete explanation for the various behaviors. The video [1] by George Hart is excellent. In addition to performing the experiment on an actual bicycle, he also shows how the result can be different for a specially constructed bicycle with a very low gear ratio. Throughout this note we assume a 1:1 gear ratio, but the methods apply to any gear ratio.

Arnold's use of "pulled back" disguises the fact that there are two ways to do it. One can literally move the pedal back, using one's hand or a string. Or one can apply a force to the pedal (such as the impulse of a hammer, or a strong wind focused only on the pedal). The force interpretation is more general, and a solution has to show first that the force causes the pedal to move backward relative to the ground; then one has to show that this backward motion yields counterclockwise rotation. The laws of physics imply that a leftward force causes leftward motion of the pedal. Using a weight on a wire that pulls the pedal left by means of a pulley, one can arrange that the force in question is gravity. If the force caused rightward motion of the pedal, then that would raise the weight, violating the law of conservation of energy. So the pedal moves left relative to the ground.

Next, we follow Hart's explanation to prove that backward motion of the pedal relative to the ground leads to counterclockwise rotation. With radius 1 for the wheel and a pedal radius $r < 1$, the motion of the pedal as the bike rolls forward follows a *curtate trochoid*, a variation on the classic cycloid (Figure 3). For a translational rightward speed of 1, the pedal motion decomposes into a translation $(t, 1)$ and a rotation $-r(\sin t, \cos t)$; at the lowest point ($t = 0$), the backward speed due to rotation is r, while the translational forward speed is 1. So the pedal moves forward at speed $1 - r$. This means that if the pedal moves backward, the bike must move backward. Arnold pulled his pedal backward, so the bike goes backward and the pedal moves counterclockwise.

If $r > 1$, then the pedal moves on a prolate cycloid, which is the path followed by a point on the bottom of a train wheel (Figure 4) because the bottom of the wheel extends below the track. For a train, the point on the edge of the wheel travels backward as the train travels forward; it is interesting that these two classic puzzles, bicycle and train, are closely related.

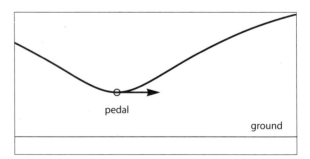

FIGURE 3. The pedal's velocity always has a positive horizontal component relative to the ground.

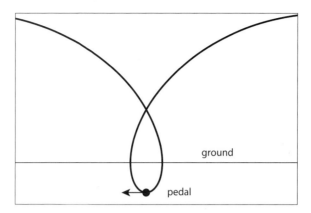

FIGURE 4. When the pedal length extends below the ground (as essentially happens with a train wheel), the pedal's motion at its lowest point is opposite to that of the bicycle.

One can ask the same question for arbitrary pedal positions. When the pedal is at the top and a force is applied in the clockwise direction, the pedal moves to the right (by the argument above), and so the bike moves forward. Therefore, there must be two pedal positions at which a tangential pedal force yields no motion at all. To understand where these static cases occur, we will consider a unicycle; the bicycle analysis is essentially the same, but there are complications because of the ratio of the rear wheel's radius to r, and also the use of gear ratios not equal to 1. Measure θ counterclockwise from the bottom (*see* Figure 5). Call the value of θ where this happens, with $0 \leq \theta \leq \pi/2$, the *static angle*. Indeed, the torque on the wheel relative to the point of contact is zero, and thus the angular acceleration around the contact point is zero.

Suppose the wheel has radius 1, F is the applied force, and f is the force of friction. Let m be the mass of the wheel, a its horizontal acceleration, I its moment of inertia (the pedal is assumed to be massless), and α the angular acceleration (positive means counterclockwise).

The rotational analogue of Newton's $F = ma$ law is $T = I\alpha$, where T is torque. So for there to be no rotation, we must have $\alpha = 0$, and because $T = f - Fr$, this means that $f = Fr$. Now, Newton's law for horizontal motion is $ma = f - F \cos \theta = Fr - F \cos \theta$, and zero acceleration a means that $r = \cos \theta$. So $\theta = \cos^{-1} r$ is the static angle at which no motion starting from rest happens, confirming what was derived earlier.

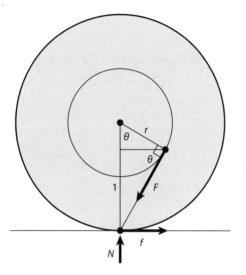

FIGURE 5. When the line perpendicular to the pedal contains the wheel's lowest point, then for typical r values, no amount of force perpendicular to the pedal arm will cause the wheel to move.

Now the following question arises: Suppose the pedal is at the static angle, and we put a huge amount of force on the pedal. Something has to give. And when it does, will the unicycle go forward or backward? The surprising answer is that if the pedal length is less than a critical radius of about 0.668, then the unicycle will not move, regardless of the amount of applied force. At some point, the wheel will crumple. If the pedal length is longer than the critical radius, then the wheel will start to slip backward, and the pedal will rotate clockwise.

The reason for this has to do with how friction works and the fact that the vertical component of the applied force causes additional friction. The coefficient of static friction (μ) for rubber and asphalt is about 0.9, which means that f_{max}, the maximum frictional force, is $0.9N$, where N is the normal force of the ground. Let W be the weight of the wheel: $W = mg$, where g is the force of gravity.

Suppose the pedal is at the static angle (*see* Figure 5, where $r = 1/2$ and $\theta = 60°$), and F is so large that the horizontal component of F exceeds f_{max}, forcing leftward motion. Then $f = f_{max}$ and $F \cos \theta > f$. From Newton's law, we have $ma = f - F \cos \theta < 0$, and so the wheel moves backward (to the left). From the torque law, we have $I\alpha = f - Fr = f - F \cos \theta < 0$, and so α is negative, and the rotation is clockwise. So there is slippage (leftward motion, clockwise rotation) when the static friction is exceeded.

But it is not so simple. It is generally impossible for f_{max} to be exceeded regardless of the size of F. For if the maximal frictional force is exceeded, we would have $F \cos \theta > f_{max} = \mu(W + F \sin \theta)$. This happens if and only if $F > \mu W/(\cos(\theta) - \mu \sin(\theta))$, and so the critical condition is $\cot \theta > \mu$. So for maximum friction to be exceeded, we must have $r/\sqrt{1 - r^2} = \cot(\cos^{-1} r) > \mu$, which is equivalent to $r > \mu/\sqrt{1 + \mu^2}$. For rubber on asphalt, μ is about 0.9, which gives $r > 0.668$. So if the pedal radius is longer than usual, the maximal friction can be exceeded and motion can be made to happen.

However, a typical pedal length of a unicycle is about 0.33, and so no amount of tangential pedal force at the critical angle θ will cause the wheel or the pedal to move. The situation for a bike is similar, but the analysis is more complicated. Try it, and you will discover the static angle for your bicycle and see that no amount of force at that angle will cause any motion.

A detailed analysis of the bicycle paradox, with some history, can be found in [4]. An interactive demonstration that allows one to vary several parameters is available at [2].

Acknowledgments

The author is grateful to Mark Levi for detailed discussions and for contributing the proof involving f_{max}.

References

1. G. Hart. Mathematical impressions: The bicycle pulling puzzle. Video available at https://www.simonsfoundation.org/2014/03/20/mathematical-impressions-the-bicycle-pulling-puzzle, 2014.

2. B. Kreka and S. Wagon. The bicycle paradox. Available at http://demonstrations.wolfram.com/TheBicycle-Paradox/. Wolfram Demonstrations Project, 2010.

3. M. Levi. Review of *Mathematical Understanding of Nature: Essays on Physical Phenomena and Their Understanding by Mathematicians*, by V. I. Arnold. *Mathematical Intelligencer* 42:1 (2020), 84–86.

4. J. D. Nightingale. Which way will the bike move? *Physics Teacher* 31:4 (1993), 244–245.

Tricolor Pyramids

JACOB SIEHLER

To tricolor a pyramid of hexagonal cells, assign each cell one of the colors red, yellow, or green according to a single rule: Each cell taken together with the two cells *directly below it* must be either (1) all three the same color, or (2) all three different colors. The coloring on the left in Figure 1 is correct, while the one on the right is not quite valid, as demonstrated by the marked cells.

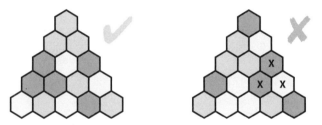

FIGURE 1. Valid and invalid tricolorings of the five-row pyramid. See also color insert.

Figure 2 has four puzzles to try as a warm-up followed by two challenges. Each one has a unique red/yellow/green coloring that satisfies the single rule. Once you have completed the warm-up puzzles, try your hand at one of the two nine-row puzzles.

A

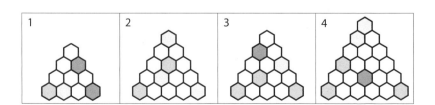

B

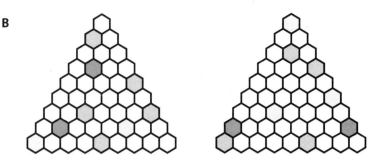

FIGURE 2. Four warm-up puzzles and two challenges.

If you find these examples interesting and want more to solve, you will find many more at http://homepages.gac.edu/~jsiehler/games /pyramids-start.html. As with similar logic puzzles, the puzzle designer must place clues carefully to create a puzzle that is not too hard and especially not too easy. These puzzles, and those online, have been chosen with care. While computer assistance is helpful in creating puzzles, understanding the underlying mathematics before jumping into computation can make the puzzles more interesting.

Consider some basic questions that a would-be puzzle designer should be able to answer: what is the minimum number of clues required to uniquely determine the solution to a puzzle, and when does a given set of clues suffice to determine a solution at all? Puzzle solvers also want to know if there are methods to find solutions without guesswork and backtracking. At first glance, it might not even be clear what branch of math would help to answer these questions: logic? graph theory? combinatorics?

Colors to Numbers

In truth, this is an algebra puzzle arising from the simplest linear equation in three variables, namely $a + b + c = 0$. Suppose that our number system is not the real numbers, but $\mathbb{F}_3$, the field of integers modulo 3. In $\mathbb{F}_3$, the only numbers are 0, 1, and 2. To add or multiply numbers in $\mathbb{F}_3$, we add or multiply as usual but reduce the result to its remainder upon dividing by 3. Table 1 shows the addition and multiplication tables.

TABLE 1. Addition and multiplication modulo 3. See also color insert.

+	0	1	2		×	0	1	2
0	0	1	2		0	0	0	0
1	1	2	0		1	0	1	2
2	2	0	1		2	0	2	1

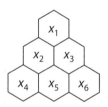

FIGURE 3. Variable assignment for three-row pyramid.

In this number system, there are nine solutions to the equation $a + b + c = 0$: three where a, b, and c all take the same value in $\mathbb{F}_3$, and six where a, b, and c take all three different values in $\mathbb{F}_3$. Thus, this equation perfectly encodes the "all three the same or all three different" condition of the puzzle. We simply replace colors with integers modulo 3—any assignment of the three colors to the three numbers 0, 1, and 2 will do. Here, we use 0 for red, 1 for green, and 2 for yellow.

Now, we can view the coloring rule for the pyramid as a system of linear equations over $\mathbb{F}_3$. For example, a three-row pyramid can have its cells labeled with variables, as in Figure 3.

Using the six variables, the coloring condition becomes the following system of three linear equations, which can be encoded into the matrix A shown to the right:

$$
\begin{aligned}
x_1 + x_2 + x_3 &= 0 \\
x_2 + x_4 + x_5 &= 0 \\
x_3 + x_5 + x_6 &= 0
\end{aligned}
\qquad
A = \begin{pmatrix} 1 & 1 & 1 & 0 & 0 & 0 \\ 0 & 1 & 0 & 1 & 1 & 0 \\ 0 & 0 & 1 & 0 & 1 & 1 \end{pmatrix}
$$

Each column of A represents a variable, and each row represents an equation. The entries are given by the coefficient of the corresponding variable in the appropriate equation.

In general, the coloring condition for a pyramid with n rows translates to a system of $n(n-1)/2$ linear equations, each with three variables summing to zero in $\mathbb{F}_3$. The solutions to the linear system are precisely the valid colorings of the pyramid. This is an elegant representation of the problem, and it allows us to study colorings systematically using the techniques of linear algebra. For example, beginning with the matrix A for the system of equations above, there is a standard linear algebra algorithm that produces the matrix

$$N_3 = \begin{pmatrix} 1 & 0 & 2 & 0 & 0 & 1 \\ 2 & 2 & 2 & 0 & 1 & 0 \\ 1 & 2 & 0 & 1 & 0 & 0 \end{pmatrix}$$

The matrix N_3 has interesting properties related to the coloring of the three-row pyramid.

1. Each row of N_3 yields a valid coloring of the three-row pyramid. For example, we interpret row 1 in N_3, which is $[1,0,2,0,0,1]$, as $x_1 = 1$, $x_2 = 0$, $x_3 = 2$, $x_4 = 0$, $x_5 = 0$, and $x_6 = 1$. Figure 4 depicts the corresponding coloring.

2. Every valid coloring of the three-row pyramid can be obtained by adding rows of N_3 together, with repetition allowed. Row addition is performed component-wise, adding numbers in the same columns (modulo 3, as usual). For example, row 2 plus row 3 plus row 3—more compactly, we will write $R_2 + 2R_3$—results in $[1,0,2,2,1,0]$, shown in Figure 5. (Try it yourself: What combination of rows makes the all-green coloring?)

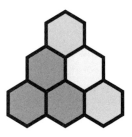

FIGURE 4. Row 1 of N_3 as a three-row pyramid coloring. See also color insert.

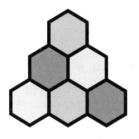

FIGURE 5. $R_2 + 2R_3$ from N_3 as a coloring. See also color insert.

3. The rows of N_3 are *independent* of one another in a strong sense. Not only are all the rows different, but no row can be written as a sum of other rows. You can prove this by considering the entries in the last three columns.

We never need to use a row more than twice in a combination because we use arithmetic modulo 3. For that reason, solutions to the three-row puzzle can be represented in a meaningful way by ordered triples of numbers from $\mathbb{F}_3$—for example, we use $(0,1,2)$ to represent the combination $0R_1 + 1R_2 + 2R_3$ and $(2,2,1)$ to represent the combination $2R_1 + 2R_2 + R_3$. The all-red (all-zero) coloring corresponds to $(0,0,0)$. With a computer, or even by hand, we could step through all of the $3^3 = 27$ triples, from $(0,0,0)$ to $(2,2,2)$, and produce the corresponding colorings by adding the appropriate rows of N_3.

In a linear algebra class, we would say that the solutions to the three-row puzzle form a three-dimensional vector space over $\mathbb{F}_3$, and the three special properties of N_3 amount to saying that the rows of this matrix form a *basis* for that space. The solution set may also be called the *null space* associated with the system of equations, and in fact, I used the NullSpace command in *Mathematica* to compute N_3, although it is straightforward to compute by hand. Almost any computer algebra system has a command for this job.

For a pyramid with n rows, applying NullSpace to the appropriate system of equations produces a matrix with n rows and one column for each cell in the pyramid. Each combination of rows added together yields a valid coloring, and every valid coloring can be produced by adding together various combinations of the rows, from $(0,0,\ldots,0)$ to $(2,2,\ldots,2)$. Therefore, the pyramid with n rows will have 3^n valid colorings; can you find a simpler, more direct proof of this fact?

How Many Clues, and Which Ones?

In the world of Sudoku, many CPU hours have been expended to find the minimum number of clues that determine a unique solution. For pyramid puzzles, the situation is simpler, and one linear algebraic consequence is the following.

Fact. It takes at least n clues to determine a unique solution for an n-row pyramid.

However, not every set of n clues will determine a unique solution for the pyramid with n rows. The clue set on the left in Figure 6 determines a unique solution, but the set on the right does not. Try it and see! The notion of independence allows us to algebraically detect the difference between the clue sets $\{1,5,8,10\}$ and $\{1,5,7,10\}$. Using the NullSpace command on the four-row pyramid system of equations produces the matrix

$$N_4 = \begin{pmatrix} 2 & 0 & 1 & 0 & 0 & 2 & 0 & 0 & 0 & 1 \\ 0 & 1 & 2 & 0 & 2 & 2 & 0 & 0 & 1 & 0 \\ 0 & 2 & 1 & 2 & 2 & 0 & 0 & 1 & 0 & 0 \\ 2 & 1 & 0 & 2 & 0 & 0 & 1 & 0 & 0 & 0 \end{pmatrix}$$

We construct two smaller matrices by picking out only those columns of N_4 that correspond to the cells in the clue sets from Figure 5: P_1 uses columns 1, 5, 8, and 10 from N_4, while P_2 uses columns 1, 5, 7, and 10.

$$P_1 = \begin{pmatrix} 2 & 0 & 0 & 1 \\ 0 & 2 & 0 & 0 \\ 0 & 2 & 1 & 0 \\ 2 & 0 & 0 & 0 \end{pmatrix} \quad \text{and} \quad P_2 = \begin{pmatrix} 2 & 0 & 0 & 1 \\ 0 & 2 & 0 & 0 \\ 0 & 2 & 0 & 0 \\ 2 & 0 & 1 & 0 \end{pmatrix}$$

Remember, P_1 corresponds to the "good" clue set that determines a unique solution, and you can see that its rows have that wonderful independence property: No row of P_1 is equal to any other row, or even to any sum of other rows. On the other hand, P_2 lacks the independence property, as rows 2 and 3 are identical.

We do not prove it here, but this process is exactly how to test whether a set of clue cells is a "good" set that determines a unique solution: Select the columns from the NullSpace matrix that correspond to your clue

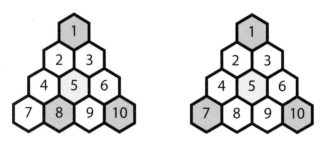

FIGURE 6. Some clues determine a unique solution, and some do not. See also color insert.

cells, and delete the rest. If the shortened rows are independent of one another, the clue set is good, but if any row can be obtained from the sum of the other rows, then the clue set allows more than one solution.

Unexpected Relations

You probably realize now that solving a puzzle can be reduced to solving a system of linear equations, which a computer can do in an eyeblink. This does not really reflect how humans solve these puzzles in practice. There are computer algorithms that are guaranteed to solve any Sudoku puzzle, but they are not practical or fun algorithms for humans. The same is true for tricolor pyramids, so let us say a little about how to design puzzles so that they have more human interest to them.

If you examine any correctly colored pyramid with four rows, you will find that the three corners of the pyramid always form a trio: all three are the same, or all three are different, regardless of how the other cells may be colored. We can explain this algebraically by assigning variables to the cells, as shown in Figure 7. Working from the bottom up (remembering that addition and multiplication are performed modulo 3, so that $1 + 2 = 0$), we see

$$x_4 = 2x_7 + 2x_8, \ x_5 = 2x_8 + 2x_9, \text{ and } x_6 = 2x_9 + 2x_{10}$$

which forces

$$x_2 = x_7 + 2x_8 + x_9 \text{ and } x_3 = x_8 + 2x_9 + x_{10}$$

so that, finally,

$$x_1 = 2x_7 + 2x_{10}$$

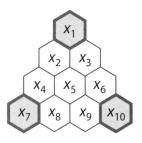

FIGURE 7. The "three corners" pattern.

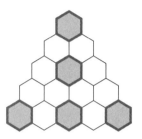

FIGURE 8. The "five-T" pattern.

or equivalently, $x_1 + x_7 + x_{10} = 0$. This means that knowledge of any two of the corners in a four-row pyramid allows the puzzler to deduce the last; the same reasoning applies to any four-row subpyramid in a larger puzzle. In warm-up puzzles 3 and 4 from Figure 2, the "three corners" pattern can be used to deduce a cell with no guesswork. In puzzle 3, the rest of the puzzle unravels quickly; puzzle 4 does not yield quite so easily.

Figure 8 shows five cells that also satisfy a simple relation among themselves, regardless of the rest of the pyramid. You will have to deduce the relation yourself this time. You can also rotate Figure 8 to find other sets of five cells satisfying identical relations. See if you can spot how to apply this pattern to solve the rest of puzzle 4 from Figure 2 with no trial and error.

Figure 9 shows a further sampling of patterns that can be used to eliminate guesswork in puzzle solving. Any one of the shaded cells in a pattern can be deduced if you know the others. Keep your eye out for these shapes (and their reflected and rotated forms) as you solve. Or, if you like the challenge of designing puzzles, see how many patterns you can weave into the solution of your puzzle.

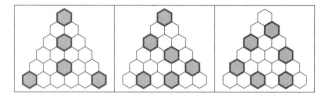

FIGURE 9. The "Y," "S," and "O" patterns.

Let us consider one final problem. How can you adapt the linear algebraic techniques described in the previous sections to determine if a particular set of cells forms a pattern? You can bet it is all about that independence property.

Final Remarks

The tricolor puzzle is a close relative of "Number Pyramid" puzzles commonly used in early grades to practice addition, subtraction, and algebraic thinking. I heartily recommend Jan Hendrik Müller's article, "Exploring Number-Pyramids in the Secondary Schools" (*The Teaching of Mathematics*, VI(3), 2003, pp 37–48; available at http://www.teaching.math.rs/vol/tm613.pdf) for a thoughtful and creative look at the mathematics of these puzzles and how they can be used in the classroom.

In the tricolor puzzles, simple patterns like "three corners" and "five-T" emerge as a consequence of reducing the coefficients modulo 3. Consequently, the puzzles acquire a geometric, pattern-spotting element that is not present in ordinary number pyramids. Nonetheless, standard algorithms and theorems from linear algebra can still be used to solve puzzles, test clue sets for solvability, search for useful patterns, and design interesting puzzles around them. Linear algebra turns out to be pretty colorful!

Does Time Really Flow?
New Clues Come from a
Century-Old Approach to Math

NATALIE WOLCHOVER

Strangely, although we feel as if we sweep through time on the knife edge between the fixed past and the open future, that edge—the present—appears nowhere in the existing laws of physics.

In Albert Einstein's theory of relativity, for example, time is woven together with the three dimensions of space, forming a bendy, four-dimensional space-time continuum—a "block universe" encompassing the entire past, present, and future. Einstein's equations portray everything in the block universe as decided from the beginning; the initial conditions of the cosmos determine what comes later, and surprises do not occur—they only seem to. "For us believing physicists," Einstein wrote in 1955, weeks before his death, "the distinction between past, present, and future is only a stubbornly persistent illusion."

The timeless, predetermined view of reality held by Einstein remains popular today. "The majority of physicists believe in the block-universe view, because it is predicted by general relativity," said Marina Cortês, a cosmologist at the University of Lisbon.

However, she said, "if somebody is called on to reflect a bit more deeply about what the block universe means, they start to question and waver on the implications."

Physicists who think carefully about time point to troubles posed by quantum mechanics, the laws describing the probabilistic behavior of particles. At the quantum scale, irreversible changes occur that distinguish the past from the future: A particle maintains simultaneous quantum states until you measure it, at which point the particle adopts one of the states. Mysteriously, individual measurement outcomes are random and unpredictable, even as particle behavior collectively follows statistical patterns. This apparent inconsistency between the nature of time in quantum mechanics and the way it functions in relativity has created uncertainty and confusion.

Recently, the Swiss physicist Nicolas Gisin has published four papers that attempt to dispel the fog surrounding time in physics. As Gisin sees it, the problem all along has been mathematical. Gisin argues that time in general and the time we call the present are easily expressed in a century-old mathematical language called intuitionist mathematics, which rejects the existence of numbers with infinitely many digits. When intuitionist math is used to describe the evolution of physical systems, it makes clear, according to Gisin, that "time really passes and new information is created." Moreover, with this formalism, the strict determinism implied by Einstein's equations gives way to a quantum-like unpredictability. If numbers are finite and limited in their precision, then nature itself is inherently imprecise, and thus unpredictable.

Physicists are still digesting Gisin's work—it is not often that someone tries to reformulate the laws of physics in a new mathematical language—but many of those who have engaged with his arguments think they could potentially bridge the conceptual divide between the determinism of general relativity and the inherent randomness at the quantum scale.

"I found it intriguing," said Nicole Yunger Halpern, a quantum information scientist at Harvard University, responding to Gisin's recent article in *Nature Physics*.[1] "I'm open to giving intuitionist mathematics a shot."

Cortês called Gisin's approach "extremely interesting" and "shocking and provocative" in its implications. "It's really a very interesting formalism that is addressing this problem of finite precision in nature," she said.

Gisin said it is important to formulate laws of physics that cast the future as open and the present as very real, because that is what we experience. "I am a physicist who has my feet on the ground," he said. "Time passes; we all know that."

Information and Time

Gisin, 67, is primarily an experimenter. He runs a lab at the University of Geneva that has performed groundbreaking experiments in quantum communication and quantum cryptography. But he is also the rare crossover physicist who is known for important theoretical insights, especially ones involving quantum chance and nonlocality.

On Sunday mornings, in lieu of church, Gisin makes a habit of sitting quietly in his chair at home with a mug of oolong tea and contemplating deep conceptual puzzles. It was on a Sunday about two and a half years ago that he realized that the deterministic picture of time in Einstein's theory and the rest of "classical" physics implicitly assumes the existence of infinite information.

Consider the weather. Because it is chaotic, or highly sensitive to small differences, we cannot predict exactly what the weather will be a week from now. But because it is a classical system, textbooks tell us that we could, in principle, predict the weather a week on, if only we could measure every cloud, gust of wind, and butterfly's wing precisely enough. It is our own fault we cannot gauge conditions with enough decimal digits of detail to extrapolate forward and make perfectly accurate forecasts, because the actual physics of weather unfolds like clockwork.

Now expand this idea to the entire universe. In a predetermined world in which time only seems to unfold, exactly what will happen for all time actually had to be set from the start, with the initial state of every single particle encoded with infinitely many digits of precision. Otherwise, there would be a time in the far future when the clockwork universe itself would break down.

But information is physical. Modern research shows it requires energy and occupies space. Any volume of space is known to have a finite information capacity (with the densest possible information storage happening inside black holes). The universe's initial conditions would, Gisin realized, require far too much information crammed into too little space. "A real number with infinite digits can't be physically relevant," he said. The block universe, which implicitly assumes the existence of infinite information, must fall apart.

He sought a new way of describing time in physics that did not presume infinitely precise knowledge of the initial conditions.

The Logic of Time

The modern acceptance that there exists a continuum of real numbers, most with infinitely many digits after the decimal point, carries little trace of the vitriolic debate over the question in the first decades of the twentieth century. David Hilbert, the great German mathematician, espoused the now-standard view that real numbers exist and can be manipulated as completed entities. Opposed to this notion were mathematical "intuitionists," led by the acclaimed Dutch topologist L.E.J. Brouwer, who saw mathematics as a construct. Brouwer insisted that numbers must be constructible, their digits calculated or chosen or randomly determined one at a time. Numbers are finite, said Brouwer, and they are also processes: They can become ever more exact as more digits reveal themselves in what he called a choice sequence, a function for producing values with greater and greater precision.

By grounding mathematics in what can be constructed, intuitionism has far-reaching consequences for the practice of math, and for determining which statements can be deemed true. The most radical departure from standard math is that the law of excluded middle, a vaunted principle since the time of Aristotle, does not hold. The law of excluded middle says that either a proposition is true, or its negation is true—a clear set of alternatives that offers a powerful mode of inference. But in Brouwer's framework, statements about numbers might be neither true nor false at a given time, since the number's exact value has not yet revealed itself.

There is no difference from standard math when it comes to numbers like 4, or ½, or pi, the ratio of a circle's circumference to its diameter. Even though pi is irrational, with no finite decimal expansion, there is an algorithm for generating its decimal expansion, making pi just as determinate as a number like ½. But consider another number x that is in the ballpark of ½.

Say the value of x is 0.4999, where further digits unfurl in a choice sequence. Maybe the sequence of 9s will continue forever, in which case x converges to exactly ½. (This fact, that 0.4999... = 0.5, is true in standard math as well, since x differs from ½ by less than any finite difference.)

But if at some future point in the sequence, a digit other than 9 crops up—if, say, the value of x becomes 4.999999999999997...—then no matter what happens after that, x is less than ½. But before that happens, when all we know is 0.4999, "we don't know whether or not a digit other than 9 will ever show up," explained Carl Posy, a philosopher of mathematics at the Hebrew University of Jerusalem and a leading expert on intuitionist math. "At the time we consider this x, we cannot say that x is less than ½, nor can we say that x equals ½." The proposition "x is equal to ½" is not true, and neither is its negation. The law of excluded middle does not hold.

Moreover, the continuum cannot be cleanly divided into two parts consisting of all numbers less than ½ and all those greater than or equal to ½. "If you try to cut the continuum in half, this number x is going to stick to the knife, and it won't be on the left or on the right," said Posy. "The continuum is viscous; it's sticky."

Hilbert compared the removal of the law of excluded middle from math to "prohibiting the boxer the use of his fists," since the principle underlies much mathematical deduction. Although Brouwer's intuitionist framework compelled and fascinated the likes of Kurt Gödel and Hermann Weyl, standard math, with its real numbers, dominates because of ease of use.

The Unfolding of Time

Gisin first encountered intuitionist math at a meeting last May attended by Posy. When the two got to talking, Gisin quickly saw a connection between the unspooling decimal digits of numbers in this mathematical framework and the physical notion of time in the universe. Materializing digits seemed to naturally correspond to the sequence of moments defining the present, when the uncertain future becomes concrete reality. The lack of the law of excluded middle is akin to indeterministic propositions about the future.

In a work published in December 2019 in *Physical Review A*, Gisin and his collaborator Flavio Del Santo used intuitionist math to formulate an alternative version of classical mechanics, one that makes the same predictions as the standard equations but casts events as indeterministic—creating a picture of a universe where the unexpected happens and time unfolds.[2]

It is a bit like the weather. Recall that we cannot precisely forecast the weather because we do not know the initial conditions of every atom on Earth to infinite precision. But in Gisin's indeterministic version of the story, those exact numbers never existed. Intuitionist math captures this: The digits that specify the weather's state ever more precisely, and dictate its evolution ever further into the future, are chosen in real time as that future unfolds in a choice sequence. Renato Renner, a quantum physicist at the Swiss Federal Institute of Technology in Zurich, said Gisin's arguments "point in the direction that deterministic predictions are fundamentally impossible in general."

In other words, the world is indeterministic; the future is open. Time, Gisin said, "is not unfolding like a movie in the cinema. It is really a creative unfolding. The new digits really get created as time passes."

Fay Dowker, a quantum gravity theorist at Imperial College London, said she is "very sympathetic" to Gisin's arguments, as "he is on the side of those of us who think that physics doesn't accord with our experience and therefore it's missing something." Dowker agrees that mathematical languages shape our understanding of time in physics, and that the standard Hilbertian mathematics that treats real numbers as completed entities "is certainly static. It has this character of being timeless, and that definitely is a limitation to us as physicists if we're trying to incorporate something that's as dynamic as our experience of the passage of time."

For physicists such as Dowker who are interested in the connections between gravity and quantum mechanics, one of the most important implications of this new view of time is how it begins to bridge what have long been thought of as two mutually incompatible views of the world. "One of the implications it has for me," said Renner, "is that classical mechanics is in some ways closer to quantum mechanics than we thought."

Quantum Uncertainty and Time

If physicists are going to solve the mystery of time, they have to grapple not just with the space-time continuum of Einstein, but also with the knowledge that the universe is fundamentally quantum, ruled by chance and uncertainty. Quantum theory paints a very different picture of time than Einstein's theory. "Our two big theories on physics, quantum theory and general relativity, make different statements," said

Renner. He and several other physicists said this inconsistency underlies the struggle to find a quantum theory of gravity—a description of the quantum origin of space-time—and to understand why the Big Bang happened. "If I look at where we have paradoxes and what problems we have, in the end they always boil down to this notion of time."

Time in quantum mechanics is rigid, not bendy and intertwined with the dimensions of space as in relativity. Furthermore, measurements of quantum systems "make time in quantum mechanics irreversible, whereas otherwise the theory is completely reversible," said Renner. "So time plays a role in this thing that we still don't really understand."

Many physicists interpret quantum physics as telling us that the universe is indeterministic. "For Chrissakes, you have two uranium atoms: One of them decays after 500 years, and the other one decays after 1,000 years, and yet they're completely identical in every way," said Nima Arkani-Hamed, a physicist at the Institute for Advanced Study in Princeton, New Jersey. "In every meaningful sense, the universe is not deterministic."

Still, other popular interpretations of quantum mechanics, including the many-worlds interpretation, manage to keep the classical, deterministic notion of time alive. These theories cast quantum events as playing out a predetermined reality. Many-worlds theory, for instance, says each quantum measurement splits the world into multiple branches that realize every possible outcome, all of which were set in advance.

Gisin's ideas go the other way. Instead of trying to make quantum mechanics a deterministic theory, he hopes to provide a common, indeterministic language for both classical and quantum physics. But the approach departs from standard quantum mechanics in an important way.

In quantum mechanics, information can be shuffled or scrambled, but never created or destroyed. Yet if the digits of numbers defining the state of the universe grow over time as Gisin proposes, then new information is coming into being. Gisin said he "absolutely" rejects the notion that information is preserved in nature, largely because "there is clearly new information that is created during a measurement process." He added, "I'm saying that we need another way of looking at these entire ideas."

This new way of thinking about information may suggest a resolution to the black hole information paradox, which asks what happens to information swallowed by black holes. General relativity implies that information is destroyed; quantum theory says it is preserved. Hence the paradox. If a different formulation of quantum mechanics in terms of intuitionist math allows information to be created by quantum measurements, perhaps it also lets information be destroyed.

Jonathan Oppenheim, a theoretical physicist at University College London, believes information is indeed lost in black holes. He does not know if Brouwer's intuitionism will be the key to showing this, as Gisin contends, but he says there is reason to think information creation and destruction might be deeply related to time. "Information is destroyed as you go forward in time; it's not destroyed as you move through space," Oppenheim said. The dimensions that make up Einstein's block universe are very different from one another.

Along with supporting the idea of creative (and possibly destructive) time, intuitionist math also offers a novel interpretation of our conscious experience of time. Recall that in this framework, the continuum is sticky, impossible to cut in two. Gisin associates this stickiness with our sense that the present is "thick"—a substantive moment rather than a zero-width point that cleanly cleaves past from future. In standard physics, based on standard math, time is a continuous parameter that can take any value on the number line. "However," Gisin said, "if the continuum is represented by intuitionistic mathematics, then time can't be cut in two sharply." It is thick, he said, "in the same sense as honey is thick."

So far, it is just an analogy. Oppenheim said he had "a good feeling about this notion that the present is thick. I'm not sure why we have that feeling."

The Future of Time

Gisin's ideas have prompted a range of responses from other theorists, all with their own thought experiments and intuitions about time to go on.

Several experts agreed that real numbers do not seem to be physically real, and that physicists need a new formalism that does not rely on them. Ahmed Almheiri, a theoretical physicist at the Institute for Advanced Study who studies black holes and quantum gravity, said quantum mechanics "precludes the existence of the continuum." Quantum math bundles energy and other quantities into packets, which are more like whole numbers rather than a continuum. And infinite numbers become truncated inside black holes. "A black hole may seem to have a continuously infinite number of internal states, but [these get] cut off," he said, due to quantum gravitational effects. "Real numbers can't exist, because you can't hide them inside black holes. Otherwise they'd be able to hide an infinite amount of information."

Sandu Popescu, a physicist at the University of Bristol who corresponds often with Gisin, agreed with the latter's indeterministic worldview but said he is not convinced that intuitionist math is necessary. Popescu objects to the idea that digits of real numbers count as information.

Arkani-Hamed found Gisin's use of intuitionist math interesting and potentially relevant to cases such as black holes and the Big Bang, where gravity and quantum mechanics come into apparent conflict. "These questions—of numbers as finite, or fundamentally things that exist, or whether there's infinitely many digits, or the digits are made as you go on," he said, "might be related to how we should ultimately think about cosmology in situations where we don't know how to apply quantum mechanics." He too sees the need for a new mathematical language that could "liberate" physicists from infinite precision and allow them to "talk about things that are a little bit fuzzy all the time."

Gisin's ideas resonate in many corners but still need to be fleshed out. Going forward, he hopes to find a way of reformulating relativity and quantum mechanics in terms of finite, fuzzy intuitionist mathematics, as he did with classical mechanics, potentially bringing the theories closer. He has some ideas about how to approach the quantum side.

One way that infinity rears its head in quantum mechanics is in the "tail problem": Try to localize a quantum system, like an electron on the moon, and "if you do that with standard mathematics, you have to admit that an electron on the moon has a super small probability of being also detected on Earth," Gisin said. The "tail" of the mathematical function representing the particle's position "becomes exponentially small but nonzero."

But Gisin wonders, "What reality should we attribute to a super small number? Most experimentalists would say, 'Put it to zero and stop questioning.' But maybe the more theoretically oriented would say, 'OK, but there is something there according to the math.'"

"But it depends, now, which math," he continued. "Classical math, there is something. In intuitionist math, no. There is nothing." The electron is on the moon, and its chance of turning up on Earth is well and truly zero.

Since Gisin first published his work, the future has grown only more uncertain. Now every day is a kind of Sunday for him, as crisis grips the world. Away from the lab, and unable to see his granddaughters except on a screen, he plans to keep thinking, at home with his mug of tea and garden view.

Notes

1. Gisin, N. Mathematical languages shape our understanding of time in physics. *Nat. Phys.* 16, 114–116 (2020). https://doi.org/10.1038/s41567-019-0748-5.

2. Del Santo, F., and N. Gisin. Physics without determinism: Alternative interpretations of classical physics. *Phys. Rev. A* 100(6), December 2019.

The Role of History
in the Study of Mathematics

HAROLD M. EDWARDS

I believe that the following summary describes the view of the history of modern mathematics that is universally believed by mathematicians today and believed by all too many historians of mathematics as well.

Summary History: The invention of differential and integral calculus in the seventeenth century by Newton and Leibniz was an enormous advance, but it remained on a shaky logical basis throughout the eighteenth century. With Cauchy in the early nineteenth century and, more seriously, with Weierstrass later in the nineteenth century, calculus was finally given a firm and rigorous foundation. What was achieved at that time was a proper understanding of limits and, more generally, of the infinite processes of integration and differentiation.

A related development in the second half of the nineteenth century that was central to modern mathematics was Cantor's creation of his theory of infinite numbers, both cardinal and ordinal, which made it possible, for the first time, to deal in a rigorous way with infinite magnitudes. Hilbert later called the complex of ideas about infinite sets that arose from Cantor's work "the paradise that Cantor created for us."

Yet another important contribution to the modern conception of mathematics in the second half of the nineteenth century was Dedekind's pioneering work in formulating the theory of algebraic numbers in terms of set theory, beginning with the notion of an ideal as a certain kind of subset of a ring. Dedekind's great inspiration was the work of his friend Riemann, who inspired him to focus on concepts in his mathematics, not formulas or computation. This was the germ of what is now called "structural mathematics," in which the focus is on structures—groups, rings, fields, operator algebras, measure spaces, topological spaces—which are defined axiomatically in terms of set theory. The combination of these ingredients has produced the rigorous modern approach to pure mathematics that has allowed for the rapid advances on many fronts that have taken place since 1900.

Readers who are familiar with my work and my views on the philosophy of mathematics will have known as they read this sketch that my purpose would be to oppose it. I believe that historians of mathematics have been far too willing to accept this widely accepted narrative of the history of modern mathematics, and that in accepting it, they fail in what I consider to be their most important function as far as the study of mathematics itself is concerned—the function of studying and analyzing the works of the great mathematicians throughout history in order to make those works accessible to modern readers and researchers.

It is understandable that historians are inclined to be strongly influenced by the opinions of mathematicians when it comes to interpreting the great metamorphosis in the understanding of the foundations of mathematics that has taken place since the time of Cantor and Dedekind, but mathematicians tend to be unreliable sources when it comes to the history of mathematics. The natural inclination of mathematicians is to work out ideas and narratives for themselves, rather than to study carefully what someone else has done. When they do study what someone else has done, it is usually to compare it to what they themselves have done, rather than to study it on its own terms, and the framework within which modern mathematicians do their thinking is ill suited to understanding many of the great works of mathematics from the past.

An outstanding example of what I mean here is given by Hilbert's famous work on the theory of algebraic number fields, commonly known as the *Zahlbericht*. This is not a historical work. As Hilbert says in the preface, his goal was to assemble the facts (*Tatsachen*) of algebraic number theory and organize them into a logical development from a unified point of view in order to "bring closer the time when the achievements of our great classics in number theory will become the common property of all mathematicians." This may appear at first to agree with the goal I state above of making classic texts available to contemporary readers, but it is in fact the very opposite. As the rest of the preface makes quite clear, Hilbert's goal is not at all to make accessible to his readers the works of Euler, Gauss, Kummer, and the others he mentions. His goal is plainly to make access to these authors unnecessary. Rather than *studying* previous authors, his goal is to *replace* them.

Abel wrote, "It appears to me that if one wants to make progress in the study of mathematics one should study the masters and not the pupils." I agree entirely, and I see this point of view as being diametrically opposed to Hilbert's point of view in the *Zahlbericht*. Hilbert says, in effect, "The masters of number theory are Gauss, Kummer, et al., and I have composed a work that arranges their achievements in a coherent and logical way that will make it unnecessary for you to study their works." Don't read the works of the masters, read the work of their pupil Hilbert, who will make their ideas easily accessible to you!

I once participated in a conference of historians of mathematics on the topic "Disciplines and Styles in Pure Mathematics." I was pleased with the word "style," because I believe that the role of style—even of fashion—in mathematics is not sufficiently appreciated. To my disappointment, I found that the other participants had such a different interpretation of the importance and meaning of "style" in pure mathematics that when they did discuss this aspect of mathematical presentation, it was only to interpret it in the very narrow sense of expository style, about which no one had anything very enlightening to say. The basic attitude seemed to be that everyone knows what the correct style for the presentation of pure mathematics is: a set of axioms regarding some type of objects in Cantor's paradise, followed by the deduction, from the axioms, of theorems about them.

The issue of "style" is enormously relevant in the case of the *Zahlbericht*. Hilbert himself would surely not have denied that the style of the classical writers whom he cites was utterly different from his own style in the *Zahlbericht*. Notwithstanding the extreme difference in style, his contention was that the facts—the *Tatsache*—were the same, so much so that readers who studied the *Zahlbericht* would gain such a complete grasp of the theory that they would come into the possession of their heritage from the masters without having to consult the masters at all. This contention—that the style can be radically different but the content the same—seems to me absurd, and I believe that a comparison of, for example, the works of Kummer to Hilbert's exposition of the theory of what he calls *Kummersche Zahlkörpern* will show that the difference in style makes for a substantial difference in the way that the "facts" are to be understood. For example, Kummer rarely strays from statements that are susceptible to verification by concrete calculation in simple cases, while Hilbert is altogether unconcerned with calculation. In his preface, Hilbert says that he sought in the *Zahlbericht* to "avoid Kummer's great calculational apparatus,"[1] which is an example of the sort of thing Abel could well have had in mind when he recommended reading the masters rather than the pupils. Kummer's "great calculational apparatus" shows us how he understood the problems he was grappling with and in this way shows how he was able to solve them so effectively. Hilbert's wish to excise this aspect of Kummer's work suggests to me that he would have had little chance of duplicating Kummer's great discoveries had Kummer not made them first. For example, is it conceivable that a researcher who believes that higher mathematics deals primarily with concepts rather than computations could discover, as Kummer did, that Fermat's last theorem is true for every prime exponent $p > 2$ that does not divide the numerators of any of the first $\frac{p-3}{2}$ Bernoulli numbers? Or that this implied a proof of Fermat's last theorem for all prime exponents less than 60 other than 37 and 59?[2]

My own presentation at the meeting on style[3] argued in favor of an algorithmic style of mathematics that is certainly far closer to the mode of thought of the masters of number theory than the modern style Hilbert wanted to use to explain their work.

I cited my paper "Euler's Definition of the Derivative,"[4] in which I had contended that Euler's legendary laxity in reasoning about differentials and limits is indeed legendary, but in the negative sense that it is fiction. As I show in the paper, a careful reading of the early volumes of his major treatise on differential and integral calculus shows how firmly grounded in computational reality his calculus was—far more so than calculus textbooks are today. As I explain in the paper, Euler carries out some computations to an accuracy of 12 decimal places and makes clear that in his conception of calculus it is necessary to be assured that arbitrarily many decimal places can be obtained if they are needed.

Also in my argument in favor of an algorithmic style of mathematics I refer to a paper I called "The Algorithmic Side of Riemann's Mathematics,"[5] in which I explained why Riemann's legendary insistence that, as Hilbert states it, "proofs should be effected not by computation but solely through concepts"[6] is also a legend in the negative sense, a fiction that does not stand up to an examination of Riemann's works. I cited several places in Riemann's works where the algorithmic nature of his thinking is undeniable. Perhaps the strongest example is a formula that has come to be known as the Riemann–Siegel formula for the evaluation of $\zeta(z)$ for z on the so-called critical line. The story of this formula is one of the great examples of important historical scholarship in the study of mathematics.

By 1930 it had become clear how important and difficult the problem of determining the truth or falsity of the Riemann hypothesis was likely to be. By then, researchers had begun serious computational investigations of the zeta function, no doubt in the hope that the problem might be solved by finding a zero that was not on the critical line. Their work ran into serious difficulty in evaluating $\zeta(z)$ accurately when $z = \frac{1}{2} + it$ for large values of t. It had been known ever since the publication of Riemann's collected works that Riemann had once said he had developed a formula for effecting such evaluations, but that he had not found a form of it simple enough to publish. Carl Ludwig Siegel, already renowned as a mathematician, followed an impulse more characteristic of a historian than of a mathematician and undertook to try to find this formula in Riemann's *Nachlass* in Göttingen. To the amazement of anyone who has seen the chaos of Riemann's private notes, not only did Siegel find the sought-for formula, he was able to derive it, to clarify it to a considerable extent, and to prove its validity. Seventy years after Riemann had developed it for his own private use in computing values of $\zeta(z)$, it was far superior to the methods others had been able to develop for the same purpose during those 70 years. (To this day, it is the essential tool used in the empirical study of the zeta function.)

In his introduction to the paper presenting the Riemann–Siegel formula, Siegel mentions what he calls the "legend" that Riemann had found his results using "grand general ideas, without needing the formal tools of analysis." He ascribes that legend to Felix Klein and confidently states that it was not as widely believed in 1932 as it had been in Klein's time, but in 2019 this legend is often retold and seldom if ever questioned.

I don't think there is any need to make the case that the work of the other pre-Cantor masters of mathematics whom Hilbert mentions in his preface—Fermat, Euler, Lagrange, Legendre, Gauss, Dirichlet, Jacobi, Kummer, Kronecker—was in a style that is very different from the nonconstructive set-theoretic style of what we call modern mathematics, the style that Hilbert expected to use to render this work accessible to students. (Hilbert explicitly mentions[7] the work of Cantor, Weierstrass, and Dedekind in describing the advances in the understanding of the nature of numbers that he believed would make possible a clear and coherent exposition of the classics of number theory.)

One need not in any way deny the successes or the appeal of modern mathematics in order to conclude that this contrast between modern mathematics and the mathematics of earlier times has created a barrier between modern students of mathematics and the masters of mathematics who preceded Cantor and Dedekind. I believe it is the duty of historians of mathematics to remove, to the maximum extent possible, this barrier. And I believe this is important for reasons other than idle antiquarian purity.

If the new formulations were regarded as supplements to the largely algorithmic formulations that preceded them—especially in number theory—there could be no objection to them, but that was clearly not Hilbert's intention when he undertook to avoid what he called Kummer's great computational apparatus, nor is it the way these formulations have been regarded since Hilbert's time. Instead, they have come to be regarded as replacements of all that came before, and what came before has been painted as limited, naive, and insufficiently rigorous. But I believe that these damning assessments are not only unjustified but seriously wrong.

My studies in recent years of the works of Galois and Abel have convinced me that in important ways, these works are misunderstood by modern mathematicians, and that study of them on their own terms—in constructive and algorithmic terms—reveals important aspects of them that are invisible to readers steeped in modern attitudes. For example, I believe that Galois's marvelous construction of the splitting field of a given polynomial, which is utterly neglected and forgotten today, gives a constructive meaning to the idea of a root of a polynomial and places all of Galois's theory on a firm and algorithmic foundation that is lacking in what is called "Galois theory" today. In the case of Abel, I believe that what is now called "Abel's theorem" (usually framed as a statement about line integrals on a Riemann surface, even though it was stated in a memoir Abel presented to the Paris Academy in 1826, the year of Riemann's birth) can be stated and proved in a purely algebraic way. In fact, it is obscured and becomes complicated by extraneous issues when it is stated in terms of Riemann surfaces.

And as anyone familiar with my work knows, I believe that the neglect of the works of Kronecker—who was an older contemporary of Dedekind and Cantor, not a predecessor—has impoverished mathematics. In Kronecker's case, not only has his published work been neglected by later generations, but his unwillingness to conform to the new nonconstructive approaches to algebra and number theory has caused him to be castigated and ostracized by the mathematical community, not during his lifetime, but very severely so in the twentieth century. In the popular history of mathematics, he has become the villain of the creation myth of modern mathematics. A legend has grown up that he was vicious toward those with whom he disagreed and that his disagreement was ridiculously retrograde and without merit. Despite the persistence of this caricature of him, there seems to be no evidence that he was guilty of any crime other than that of respectfully disagreeing with the adoption by his contemporaries of methods of dealing with infinity that he regarded as illegitimate. I have looked for such evidence, but have found that those who make charges against him, when they provide any evidence at all, provide bogus evidence. The deplorable neglect of the works of this great master—so admired in his own time that he was the first choice after Riemann's death to assume the chair at Göttingen that had been occupied by Gauss, then Dirichlet, then Riemann—is the inevitable result of this pernicious legend.

Political historians use the term "Whig history" to describe a type of history that is to be frowned upon, namely, a history that describes what happened in earlier times as a series of struggles to reach the state of enlightenment and stability that has now been attained, a history that does not allow for wrong turns or legitimate differences of opinion.

What many historians of mathematics produce today is, in my opinion, a Whig history of mathematics. There is a willingness to believe that we now know the proper way to formulate mathematics and that this knowledge justifies us in looking upon the works of such masters as Euler and Abel and Galois as curiosities, not worthy of the attention of present-day mathematicians, because they were written before correct formulations and rigorous proofs were possible.

As I said before, I endorse Abel's opinion that those who wish to make progress in mathematics should study the masters. If that is going to be possible, scholars—whether they call themselves mathematicians or historians of mathematics—will have to devote more effort to analyzing and elucidating the works of the masters in their own terms and less effort to reinforcing the view that the modern style of mathematics is the only one.

Acknowledgments

This paper is a revised version of a presentation written for the 5th International Conference on the History of Modern Mathematics, held in August 2019 at Northwest University in Xi'an, China. Unfortunately, an injury prevented the author from traveling to China, and the paper was presented in absentia.

Notes

1. *Ich habe versucht, den großen rechnerischen Apparat von Kummer zu vermeiden.*
2. *See* § 137 of the *Zahlbericht* or p. 182 of my book *Fermat's Last Theorem.*
3. *See* "talks" at my website, math.nyu.edu/faculty/edwardsd/.
4. *Bulletin of the AMS* 44 (2007), 575–580.
5. *See* "articles" at my website.
6. *[D]ie Beweise nicht durch Rechnung, sondern lediglich durch Gedanken zwingen soll.*
7. Page 66 of volume 1 of Hilbert's *Gesammelte Abhandlungen.*

"All of These Political Questions": Anticommunism, Racism, and the Origin of the Notices of the American Mathematical Society

MICHAEL J. BARANY

1. Introduction: "Material of Temporary Interest"

The *Notices of the American Mathematical Society* has come a long way since its first issue in February 1954 [1, pp 121–126]. Established at the October 1953 American Mathematical Society (AMS) Council meeting as a spinoff of the *Bulletin of the AMS* for "material of temporary interest, which would in general be discarded quickly," the *Notices* grew through the American mathematical community's post–World War II ascendency in the international profession to become the world's most widely read magazine for professional mathematicians.[1] With the advent of its Letters department in the June 1958 issue, the *Notices* quickly became a favored high-profile venue for the latest hot-button questions and controversies regarding matters both internal to the mathematics profession and concerning its relation to the wider world.[2]

Since then, successive editors have tried to strike a balance between serving as an official conduit for news and information from the AMS and representing the broader interests and concerns of the American and international mathematics professions.

Well before the December 2019 issue reached mathematicians' mailboxes, the magazine's online edition had begun to cause a stir along precisely these lines. As part of a series of invited commentaries from current AMS officers, vice president Abigail Thompson weighed in on the profession's debates over how to create a diverse and inclusive discipline with a provocative comparison. Recent hiring practices in the University of California (UC) system, she asserted, were reminiscent of the notorious events of 1950 when the UC Regents demanded anti-Communist loyalty oaths of all California faculty.

Reactions were swift and divisive, prompting a special online-only supplement featuring individual and collective letters to the editor signed by hundreds of concerned readers. Mathematicians around the world debated questions of politics and professional autonomy, diversity and racism, free speech and responsibility, the role of AMS officers and their official magazine, and more. The events of 2019 did indeed look something like those of 1950, but not necessarily in the ways Thompson suggested in her commentary.

A close look at the official archives of the AMS around the time of the 1950 loyalty oath controversy shows a society grappling with precisely the kinds of concerns that animated responses to Thompson's commentary [1, pp 297–300]. This older collection of materials of temporary interest, thankfully not discarded but filed away by AMS Secretaries and their support staff, show a professional community struggling to define its purposes and limits in the early years of America's postwar Civil Rights Movement and Red Scare. The society's response to the crisis in California quickly became entangled in a complex web of issues old and new and exposed weaknesses in the society's approach to decision-making and communication in the face of urgent challenges. The legacies of these controversies, including the creation of the *Notices*, continue to be felt.

2. *"Not a Political Affair"*

The 1950 loyalty oath controversy landed in the middle of a tense and monumental period for American mathematicians and their political commitments. AMS officers led by Marshall Stone, the society's president from 1943 to 1944, had deliberately used World War II as an opportunity to court military and government sponsorship for mathematical teaching and research, effectively recognizing that certain kinds of political involvement could be necessary and even lucrative for the profession [2]. They emerged from the war in a prime position to lead the effort to rebuild an international mathematics community, including a leading (but not automatic) claim to host the first postwar International Congress of Mathematicians (ICM) and to reconstitute a defunct International Mathematical Union (IMU).

The Pacific war had barely concluded when Stone's successor as AMS President, Theophil Hildebrandt, insisted to the committee laying the groundwork for an American-hosted ICM that the prospective meeting should be "completely international," underscoring "that science is not a political affair, but international in character" [3]. But it was not at all clear what it meant to be "international" or "not political" in the war's wake. Should "notorious Nazis" [4] be allowed to participate? How would the organizers assure sufficient participation from war-devastated countries whose mathematicians barely had money for food and shelter, much less transoceanic conference travel? Stone led the charge for control of the IMU organizing process on the grounds that only the Americans could avoid the political barriers he blamed for the interwar union's demise, an argument his foreign interlocutors did not always find convincing. Part of Stone's plan, quickly embraced by the ICM organizers, was to do everything they could to avoid national, political, or related criteria for participation (*see* [5, Chapters 1, 3, 4]).

In 1948, it became clear that the greatest challenge to the Americans' commitment to internationalism would be their own government's growing frenzy of anticommunism. That March, the U.S. House of Representatives' Committee on Un-American Activities challenged the loyalty of Edward Condon, the politically liberal and internationalist director of the National Bureau of Standards, precipitating alarm across the American scientific community [6].[3] At its April Council meeting, the AMS declared its "grave concern" at Condon's treatment and its potential chilling effect on scientists in public service [8, p 629]. Adding Russian to the official languages of its upcoming International Congress at the same meeting, the Council might not yet have recognized the full extent of their impending difficulties from American anticommunist policy.

By the summer of 1949, the AMS had effectively written off the prospect of Soviet participation, but organizers soon had to confront the prospect that communist and fellow traveler mathematicians from friendly countries might be denied entry to the United States for their political beliefs. The highest profile case was that of French mathematician and former Trotskyist legislative candidate Laurent Schwartz, who was selected in the midst of his visa troubles as one of 1950's Fields Medalists [9, pp 384–389]. French mathematicians threatened to boycott, and Stone joined their call and resigned his ICM roles in protest while retaining the IMU organization as a separate endeavor. While much of the AMS debate on the California situation played out after the 1950 Congress, news of the events in California hovered in the background of harrowing last-minute diplomatic negotiations for the ICM, and accusations and recriminations from those months remained raw even after the delegates had come and gone.

3. Into the Archives

The archives of the AMS today reside in the John Hay Library of Brown University in Providence, Rhode Island, the long-deferred outgrowth of a legacy of AMS entanglement in operations and publications with the university and city (*see* [1, pp 317–325]).[4] The archives grew haphazardly in files assembled at AMS headquarters and held by successive AMS Secretaries over the years, only being systematically assembled and sorted years after the fact. As the outgrowth of assorted files used primarily to support AMS operations, their organization partially reflects how they would have been collected and consulted, as working records of letters and reports that the AMS Secretary might need to consider in ongoing society business.

Not all documents that crossed the Secretary's desk were kept. When a document appears in an institutional archive like that of the AMS, it indicates that someone considered it legally or politically or otherwise important enough that it might need to be seen again and that everyone who handled it later agreed it was worth preserving. Indeed, these records show regular evidence of cross-referencing, quotation, and reuse. When an issue required sustained attention from an AMS officer, it got its own folder for each year when it was salient. Sporadic topics requiring a durable record might be clustered together based on subject headings, though it is not generally clear whether these were organized in this way initially or only in retrospect.

The result is a skewed record of what a few people in positions of authority considered useful, worrisome, or otherwise significant. These archives show only a small portion of everything that went into official AMS operations, which themselves were just a part of the wider network of activity that helped the AMS run. Reading these files, one must always ask why they were saved and what might not have been saved, and whose views and priorities are reflected.

Consider the folder "U 1949." In a working archive, the most recent documents are typically added to the tops of folders, so to read them chronologically we must start at the back. There, with the archival heading "Un-American Activities," one finds a December 1948 letter from the Bureau on Academic Freedom of the National Council of the Arts Sciences and Professions calling for scientific associations to stand in defense of academic freedom. AMS Secretary J.R. Kline's January 1949 reply explained that the AMS Council had declined to take further specific actions, but had instructed him to share its resolution from the previous April regarding the Condon Affair. This simple exchange of letters reflects an organization's attempt to make common cause with the AMS, the AMS's polite deflection, and the Secretary's record of what had been sent and on what terms.

A slightly longer series of letters sits on top of these, filed here because they concerned the "United Forces for God Against Communism [UFGAC]." Kline had initially ignored UFGAC's October 1949 letter informing him that the AMS had been elected a "Component Member" of their "non-sectarian and non-partisan" crusade, but learned in December that UFGAC had been listing the AMS on its fundraising missives. Kline thanked his informant and promptly sent UFGAC a registered letter explaining that the AMS "does not enter into the field of political activity" and that UFGAC should not use its name. Later, the letter's delivery receipt was stapled on top to round out the file. Here, Kline needed to establish a paper trail of what he knew and communicated and when as a defense against further unauthorized use of the AMS's imprimatur.

4. *"Arbitrary and Humiliating Conditions"*

The AMS Council made its first official statement on the California situation at a meeting on September 1, during the 1950 International Congress, following a turbulent summer of petitions and reversals. Their official statement, which Kline transmitted to the UC President and Board of Regents the next day, denounced the "arbitrary and humiliating conditions" imposed on the university's faculty. This and subsequent folders concerning the California loyalty oath crisis appear in the archive under the heading "Political."

The "Political 1950" folder continues with a November update from Berkeley's mathematics chairman and a December 1 letter from a UCLA mathematician urging an AMS boycott of UC campuses as meeting sites, a position the correspondent believed was shared across mathematics departments at the major UC campuses. These letters sit next to another from November 1950 detailing the AMS's initial involvement in an American Association of University Professors (AAUP) investigation into Pennsylvania State College's dismissal of Lee Lorch, a mathematician and civil rights activist who was fired for his attempts to overturn and then to circumvent racial covenants restricting access to housing in New York City.

These two concerns reverberated in the subsequent year across two parallel headings: California in "Political 1951" and Lorch in "Discrimination 1951," each divided across multiple folders. The "Political" folder opens with a bombshell: Kline's successor Edward Begle wrote on January 3 to Berkeley's Griffith Evans that the AMS Council had voted on December 28 to adopt the proposed UC boycott but that a general vote of AMS members at the Gainesville, Florida, meeting the next day had gone in the opposite direction. Ensuing documents amplify the sense of conflict: a fiery Council report denounced the UC Regents; a disappointed AMS member discontinued his donations to the society for "its unwillingness to give even moral support to those of its members who are fighting for academic freedom" and instead directed the money to those who lost UC positions; a waffling reply by the AMS Executive Director defended the society's indecision wherever "there will be a sharp division of opinion among mathematicians"; the American Civil Liberties Union Executive Director congratulated the AMS on denouncing the loyalty oaths; a telegram urged the AMS to "continue its traditional scientific detachment from matters political." In April, Marshall Stone wrote to urge the AMS to appoint a committee to resolve the Gainesville muddle.

Meanwhile, under "Discrimination," Lorch, now appointed at the historically black Fisk University, wrote of a different set of arbitrary and humiliating conditions. Four members of his new department had been denied entry to the segregated conference dinner of the March southeastern regional meeting of the Mathematical Association of America (MAA) at Peabody College and Vanderbilt University in Nashville, Tennessee. Lorch now wrote to transmit his department's request that the AMS (as well as the MAA) adopt a formal policy against racial discrimination at its meetings. A follow-up letter added 61 signatories in support of the first and enclosed a further supporting letter.

Following the "Political" dossier into the summer, records show that the committee Stone requested had been appointed, with Stone as chair. A June 1951 memorandum summarized a developing loyalty oath crisis in Oklahoma. Another June letter shows Stone's developing thinking about California, supplying him with requested information about the AMS response (or lack thereof) to a case of a Canadian mathematician accused of espionage, the Condon Affair, Lorch's AAUP case, the dismissal of Jewish mathematicians in fascist Italy, and the "racial complications which have arisen from time to time in connection with our meetings in the South" on which "The Council has never taken any action." A subsequent letter signals Stone's awareness of the emerging Oklahoma story, and later letters show Stone continuing to draw on the AMS archives through letters to Begle. Come August, Stone's committee was ready to report.

5. *"Poles Apart"*

Or rather, Stone's three-person committee was ready to file two opposing reports. Donald Spencer, who seemed to Stone relatively unconcerned with the details of the matter, joined the committee chairman in a forceful majority report arguing that the AMS had a long history of involving itself in political matters to ensure support for mathematicians, and that "the social transformations taking place in our times tend to integrate the professional activities of mathematicians ever more closely" with systems "entrusted to politicians and administrators." Stone need not have pointed out his own past role in seeking politically entangled support on the AMS's behalf. The majority report then outlined the legal and procedural basis for the AMS Council to sanction the University of California and to act decisively in future situations. In the California case, the report argued in detail, such actions were both justified and worthwhile.

Theophil Hildebrandt was, as Stone put it, "poles apart" in his assessment. The AMS was not a labor union, and therefore had no business in employment matters, according to Hildebrandt. "An unemotional report of situations affecting members of the Society" could be valuable, but he thought decisions taken "in atmosphere of high tension with inadequate background" lacked legitimacy. Hildebrandt closed by comparing the proposal to boycott UC campuses with World War I era boycotts of music by German composers: "such action smacks of child-ishness" and "will seem absurd in the future."

Hildebrandt had an ally in Kline, who opined in September 1951 that the AMS "should not be entering upon all of these political questions" and should hold instead to "our purely scientific purposes." The two, along with two others, petitioned the Council in November to revisit its acceptance of the majority report, having not been there to oppose it in person at the September meeting. Further letters defended the decision. But with the California situation appearing headed toward a resolution and with news of further setbacks in Oklahoma making the latter situation appear especially irremediable, appetite for a confrontation on loyalty oaths waned by the year's end.

This attitude toward accommodation and limited action affected the response to the Fisk mathematicians' petition, which the Council initially addressed with a bland intention "to obtain assurance" that "there will be no discrimination" if such an assurance seemed necessary. Lorch was not satisfied, and with good reason. As Alabama Polytechnic Institute (the forerunner of Auburn University) prepared to host an AMS meeting, they elected not to hold an official banquet, but also not to list any housing or dining facilities that would accommodate patrons of color. The one Black mathematician who attended drove 20 miles each way so he could sleep at home and was told he could "technically" attend the "Social Hour" held in lieu of an official banquet but "probably would not want to do so."

Another letter in the "Discrimination" folder urged that entertainment at meetings be free to follow "the social customs and wishes" of the host, unsubtly endorsing de facto segregation. A subsequent report lamented that "to omit the formal social activities at meetings where the question of race is a problem is simply the price which may temporarily have to be paid" to continue to hold meetings in the South. As Begle summarized in a January letter in the "Discrimination 1952" folder, "Few, if any, on the Council wished to crusade against discrimination, but practically everyone felt that the proper thing to do would be to sacrifice the social functions when necessary."

By the end of 1952, facing a barrage of challenges connected in one way or another to the society's debates about politics and discrimination, AMS officers moved to extricate themselves by any mechanism available. The committee on the Oklahoma Loyalty Oath Affair recommended gently censuring the Oklahoma legislature and respective university officials for actions that, although deemed legal, had clearly damaged the standing of a well-recognized department. In November, a North Carolina mathematician wrote urgently to tell his side of a story of denied accommodation "in case a protest is registered," as indeed one was by Howard University's David Blackwell a few days later, alleging racial discrimination. Begle replied to Blackwell that the North Carolina meeting had technically met the AMS's standards for non-discrimination, to which Blackwell responded "since discriminatory housing arrangements are compatible with the present requirements, a stronger statement is needed."

6. *"A Small Periodical"*

Begle's frustration with AMS mechanisms for addressing political challenges led him to convene a Committee on Controversial Questions. One member, G. Baley Price, served as well on the editorial committee of the *Bulletin of the American Mathematical Society*. Price generally agreed with Begle's complaints and proposals, but worried that there was "no good means of communicating with the entire membership" should the need arise. In February 1952, as multiple controversial questions raged, Price wrote to "suggest that the Society establish a small periodical in which can be placed all information for members which lacks a permanent character or value." Such an inexpensive spin-off of the *Bulletin* could "solve some of our present difficulties . . . communicating with the membership on questions of current interest." Taking such fleeting communications out of the *Bulletin*, moreover, would allow changes to the *Bulletin's* production and distribution that could even save the society some money. Two years later, heralded in remarkably similar terms, came the first issue of the *Notices*.

The AMS Council at first rejected the idea of a Letters column, but the *Notices* editor ultimately introduced a Letters department in 1958. The first entry discussed ethics in mathematical publishing, a subject the letter writers noted "has not yet been covered by the otherwise comprehensive Bourbaki." The very next letter the *Notices* received, the first to run in the subsequent issue, was a report by David Blackwell on an effort to raise legal defense funds for Lee Lorch, who had been prosecuted for contempt of Congress after refusing questions about his

Communist party membership from the House Committee on Un-American Activities. Lorch himself was an early contributor, too. In 1960, he denounced a politically charged invocation of "Iron Curtain countries" in a previous issue, noting that the United States had its share of restrictions on the free movement of goods and people. Among the latter, Lorch cited Paul Erdős, whose 1952 U.S. visa troubles—still unresolved in 1960—appear alongside Price's letter in the "Political 1952" folder, and whose cheerful reports of travels abroad also pepper early *Notices* Letters sections.

The AMS archives do not offer a recipe for resolving institutional conflict, defending academic freedom, or promoting racial justice. They do, however, show the long-running links between these features of the history of American mathematics. Such links run not just conceptually and ideologically but in the personal and bureaucratic mingling of letters, reports, meetings, and files. These ties can offer nuance and complexity to efforts to draw simple lessons from the past, but they can also underscore moments of clarity, conviction, and connection that have been muddied by time.

Notes

1. The quotation is from issue 1 of the *Notices*. The American Mathematical Society has recently digitized the entire back catalogue of the *Notices*, available on the *Notices* website. The "most widely read" characterization is from the magazine's current "About" page.

2. Pitcher [1, pp 122–123] notes that controversy over responsibility for the Letters section motivated several changes to the editorial structure of the *Notices*, including the 1976 creation of an Editorial Board.

3. For a personal and historical account of mathematicians' encounters with the House Un-American Activities Committee in the wake of the Condon Affair, including other episodes discussed in this essay, *see* [7].

4. Records cited here are from boxes 36 (1950 documents, 1951 California Report), 37 (1951–52 documents), and 15 (1952 Oklahoma Report) of [10].

References

[1] Everett Pitcher. *A History of the Second Fifty Years, American Mathematical Society, 1939–1988.* AMS, 1988.

[2] Michael J Barany. The World War II origins of mathematics awareness. *Notices of the AMS*, 64(4):363–367, 2017.

[3] Hildebrandt to Morse, 21 Nov 1941, Papers of Marston Morse, Harvard Depository HUGFP 106.10, courtesy of the Harvard University Archives, Cambridge, MA, box 7, "ICM-Emergency Committee" folder.

[4] Kline memorandum enclosed in J.R. Kline to Warren Weaver, 29 Oct 1946, folder 1546, box 125, Record Group 1.1, Series 200D, Rockefeller Foundation Archives, Rockefeller Archive Center, Sleepy Hollow, NY.

[5] Michael Jeremy Barany. *Distributions in postwar mathematics*. Ph.D. thesis, Princeton University, 2016.

[6] Jessica Wang. Edward Condon and the Cold War politics of loyalty. *Physics Today*, 54(12):35–41, 2001.

[7] Chandler Davis. The purge. In *A Century of Mathematics in America, Part I*, Peter Duren, ed., pp 413–428. AMS, Providence, RI, 1988.

[8] J.R. Kline. The April Meeting in New York. *Bulletin of the American Mathematical Society*, 54:622–647, 1948.

[9] Michael J Barany, Anne-Sandrine Paumier, and Jesper Lützen. From Nancy to Copenhagen to the world: The internationalization of Laurent Schwartz and his theory of distributions. *Historia Mathematica*, 44(4):367–394, 2017.

[10] Archives of the American Mathematical Society, Ms. 75. John Hay Library, Brown University, Providence, RI.

Reasoning as a Mathematical Habit of Mind

Mike Askew

Introduction

In this paper, I look at different aspects of mathematical reasoning and argue that we need to make sure students of all ages engage in a range of mathematical reasoning, particularly given the evidence that teaching for reasoning is a powerful complement to teaching that is more focused on skills and procedures.

A major report [1] published in the United States put forward a model of mathematical competency as comprising five strands woven together (this metaphor deliberately chosen on the basis of a rope being stronger than the sum of its strands): procedural fluency, strategic competence, adaptive reasoning, conceptual understanding, and productive disposition. Since the publication of this report, there has been growing consensus, within the world of mathematics education, on the importance of at least some, if not all of these strands. Linked to this, England's 2015 National Curriculum for Primary Schools [2] has the stated aim of developing fluency, problem solving, and reasoning, and the language of these three proficiencies is increasingly also being used when talking about the curriculum for Secondary Schools.

Curriculum-supporting documents often present these proficiencies in the order in which they are stated in the curriculum document, and that, along with the popularly held view that fluency in basic arithmetic is needed before students can engage in mathematical reasoning, means that mathematical reasoning is sometimes talked about and enacted in classrooms as the icing on the cake of mathematics teaching and learning—something a few students (the mathematically talented ones) get to engage with. A result of this is that some students then experience fewer opportunities to engage in mathematical reasoning than their peers.

I argue here that an emphasis on mathematical reasoning is an educational right to which all students are entitled, that reasoning is complementary to procedural fluency, not an outgrowth of it, and that rather than being some esoteric way of thinking that only a minority of students can engage in, mathematical reasoning is achievable by the vast majority of students. In making reasoning available and accessible to most students, not only will they deepen their mathematical understanding, they may also get a stronger sense of the pleasure of mathematics, of the romance of mathematics, to paraphrase the words of the illustrious previous president of the Mathematical Association, Alfred North Whitehead in [3].

Reasoning or Arithmetic?

Before reading on, I invite you to consider whether each of the following statements is true or false.

$$39 \times 46 = 39 \times 45 + 46$$
$$39 \times 46 = 39 \times 45 + 39$$

The mathematically astute reader (as the reader of this book doubtless is!) may invoke the distributive law, reasoning that $39 \times 46 = 39 \times (45 + 1)$ leading to $39 \times 45 + 39$, thus establishing the second statement as true (and consequently that the first statement is false). Offering these equations to teachers and students, given that the numbers are sufficiently "ugly" to discourage checking by calculating each side of the equation, a conversation typically ensues as to whether to "read" 39×46 as "forty-six groups of thirty-nine" or as "thirty-nine groups of forty-six," thus engaging participants in some reasoning about the commutative nature of multiplication and when it is useful to invoke its use. The reading of "forty-six groups of thirty-nine" can lead to the awareness that if this is reduced to "forty-five groups of thirty-nine" (39×45), then to preserve the equality, another group of thirty-nine needs to be added, hence the second equation is true.

Notice that in each case, reasoning as to whether each statement is true or false is independent of any ability to carry out any of the calculations presented. Now consider the second equation within a string of similar examples:

$$3 \times 5 = 3 \times 4 + 3$$

$$39 \times 46 = 39 \times 45 + 39$$
$$326 \times 18 = 326 \times 17 + 326$$

The reasoning required to establish that the first equation is true is no different from that required to establish the truth of the second or third statements. The last equation is adapted from an item on one of England's recent national tests for the end of primary school. At the time of writing, students, on exit from primary school, sit three mathematics tests—one arithmetic test and two "reasoning" tests. The last question on the 2016 second reasoning test gave the equation:

$$5{,}542 \div 17 = 326$$

Students were asked to show how they could use this equation to find the answer to 326×18, in other words, to apply the reasoning implied in the third example in the string above. This was the most poorly answered question of all the questions across all three papers, with only 26% of students answering it correctly. Given that this was the final question across all three papers, the test setters presumably thought it was the hardest item on the test: the low success rate would appear to support this. One might expect that the low score was a consequence of many students simply not getting that far on the paper, but 79% actually attempted an answer. I suspect that many of these tried to carry out some calculation, but, if one can engage in the sort of "reading" and reasoning outlined above, then the only real challenge that the problem presents is appreciating that $5{,}542 \div 17 = 326$ informs you that $326 \times 17 = 5{,}542$, and then to reason that 326×18 must therefore be $5{,}542 + 326$.

This example gets at the heart of the distinction in [4, p. 3] that the psychologists Nunes, Bryant, Sylva, and Barros make between arithmetic ("learning how to do sums and using this knowledge to solve problems") and mathematical reasoning ("learning to reason about the underlying relations in mathematical problems they have to solve"). From a number of research studies with young children, Nunes and colleagues argue that being able to do arithmetic and being able to reason mathematically cannot be treated as proxies for each other and that mathematical reasoning needs to be attended to in its own right. They examined this by tracking students in a five-year-long longitudinal study, concluding that reasoning and arithmetical abilities contribute independently to predicting progress in learning mathematics, but that of the two, "mathematical reasoning was by far the stronger predictor" [5, p. 136] of later success and that teaching must address improving reasoning skills as well as, and separately from, arithmetical skills.

A Language of Mathematical Reasoning

If mathematical reasoning is to play a more central and distinct role in teaching and learning, then I think it helps to bring some clarity to what it might look like in classrooms. A search across the mathematics education literature brings up a range of terms involving reasoning, including additive and multiplicative reasoning, statistical reasoning, proportional reasoning, approximate reasoning, geometric reasoning, both positional and axiomatic, and so on.

More concisely, [2, p. 99] describes the aim of being able to reason in the following terms:

> reason mathematically by following a line of enquiry, conjecturing relationships and generalisations, and developing an argument, justification or proof using mathematical language.

Working with teachers, I am often asked if reasoning is not simply part of problem solving, that, if students are working on solving problems (genuine ones, that is, problems to which they do not have an immediate solution method), then surely that must involve some form of reasoning. I would agree, but reasoning needs to go beyond finding an answer to a particular problem. It needs, as in England's National Curriculum definition above, to move to thinking about generality, and so problem solving is embedded within reasoning. Reasoning is broader than finding a solution to a particular problem, since reasoning is driven by the desire to ask "What is the general mathematical structure that this particular problem is only a single instantiation of?"

If we take seeking generality as the core of mathematical reasoning, then I find it helpful to attend to different types of reasoning. Two of these, deductive and inductive reasoning, are part of the canon of mathematics and need little introduction. But three others, abductive, analogical, and relational reasoning, are perhaps less often acknowledged but are important in teaching and learning mathematics. Let us look at each type.

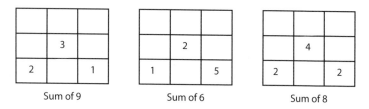

FIGURE 1: Which of these can be completed to make a magic square?

Deductive Reasoning

Magic squares are a classic example of deductive reasoning: given the sum of the rows, columns, and diagonals, and some cell entries given, can the other cells be completed? In [6], Arcavi developed an extension to the traditional magic square by allowing numbers to be repeated. So, given that a number can be used more than once, which of the squares in Figure 1 can form a magic square?

The second example introduces the use of negative numbers, while the third example raises the possibility of a solution not always being possible, opening up inquiry into the necessary conditions for being able to complete a square.

Inductive Reasoning

Take these four calculations

$$5 \times 6 =$$
$$7 \times 8 =$$
$$4 \times 5 =$$
$$6 \times 7 =$$

Presented thus, they may provide students with an opportunity to practice recall of multiplication bonds, but they offer little opportunity for mathematical engagement beyond that.

Now consider the same calculations (and their answers) in a different order:

$$4 \times 5 = 20$$
$$5 \times 6 = 30$$

$$6 \times 7 = 42$$
$$7 \times 8 = 56$$

Asking learners, "What do you notice?" and "What do you wonder?" can lead to them noticing that the answers increase by differences of 10, 12, and 14, and to wondering if this pattern will continue. By simply re-ordering the calculations, inductive reasoning is invoked (in the more everyday sense of pattern spotting, rather than in the sense of proof by induction), raising a mathematical awareness that could be explored further.

Abductive Reasoning

Largest-smallest difference is a classic example of a task that invokes abductive reasoning—noticing a repeated connection. Close to, but not quite the same, as inductive reasoning, abductive reasoning is based on noticing similarities and co-occurrences of phenomena.

Write down three digits.

1. Arrange the digits from largest to smallest as a single number.
2. Arrange the digits from smallest to largest as a single number.
3. Find the difference between the two numbers.

Repeat with the answer.
What do you notice? What do you wonder?
Here we see the complementary relationship between abductive and deductive reasoning. Trying out several examples raises the possibility that there is a generality, but establishing whether or not the generality will always hold is established by deductive reasoning. It is important that even early in their experiences of learning mathematics students are exposed to the idea that, when asked, "Will that always work?" that the answer "Well, it did the five times I tried it" is not mathematically sufficient.

Analogical Reasoning

A mathematical answer to 27 divided by 6 is 4 remainder 3.

Before reading on, I invite you to think of a simple real-world situation that can be modeled by dividing 27 by 6 but where it makes sense, back in the real world, to round the answer up to 5.

Asked to do this, people come up with situations comparable to 27 people going somewhere, booking taxis that can carry 6, and so needing to book 5 taxis, or packing all of 27 eggs into boxes of 6. Whatever the situation, it is most likely that the context chosen is some sort of "packing" situation where one "container" is not completely full, resulting in the need to round up the mathematical answer.

Research has shown that expert problem solvers do not always treat new problems from scratch, but, instead, "match" the problem analogously to other similar problems that provide a solution image or approach [7]. Here, the archetypal divide-and-round-up-the-answer analogy is to a "packing" problem. There is an extensive body of research into the power of working with "core" archetypal problems for developing understanding of additive and multiplicative structures, and I recommend anyone who is interested to read [8].

Relational Reasoning

Consider the situation below (adapted from [9]).

- On Saturday, some friends came to tea. We shared a packet of biscuits, equally.
- On Sunday, I had another tea party.
- A greater number of friends came around on Sunday.
- We shared, again equally, the same number of biscuits as there were on Saturday.

Did each person on Sunday eat more, less, or the same as each did on Saturday?

Even very young children can reason that everyone on Sunday gets less to eat (working on the assumption that these people exist in a mathematical universe where no one is on a diet, gluten intolerant, or gives up their share!). Given that there are three possible situations for the number of friends on Sunday (fewer, same, more) and for the number of biscuits (fewer, same, more), then there are nine possible differences between Saturday and Sunday, and in seven of these, it is unambiguous as to how the Sunday situation compares to Saturday.

As this example shows, reasoning about relations between quantities—relational reasoning—can be done without needing to know the actual numerical values of the quantities. Indeed, arithmetical problem-solving requires the relationship between quantities to be figured out before actually working on the calculation. Attending too quickly to the actual quantities, rather than the relationship between then, can lead to erroneous reasoning, as this example shows:

> Two battery-operated cars—one red, one blue—are traveling equally fast around a track.
> The red car started first.
> When the red car had completed 9 laps, the blue car had completed 3 laps.
> When the blue car had completed 15 laps, how many laps had the red car completed?

Offered this problem, many students close quickly to the answer of 45 laps—the various ratios between the values given make this a seductive answer—3:9 as 15:45 or 3:15 as 9:45. The correct answer of 21, given that the cars are traveling at the same speed, and so red is always 6 laps ahead of blue, cannot be established from examining the numbers, only from the information in the first sentence (often glossed over by the reader).

The Absence of Reasoning in Classrooms

Despite the evidence both for the importance of students engaging in mathematical reasoning, and for its importance in later development, much evidence points to a paucity of reasoning in mathematics lessons. Why might this be so? It is beyond the scope of this paper to explore in detail the variety of reasons for the lack of attention to reasoning (not the least of which is the climate of high-stakes testing to which schools are subjected), but two possible constraints are worth touching upon: the emotional cost of reasoning and the focus of planning.

A major research project in the United States [10] worked with a group of upper primary and middle school teachers to design and implement a number of lessons that would have mathematical reasoning at their heart. The teachers came together to discuss the mathematical tasks forming the core of the lessons, to work through the tasks, and to come to a common understanding of the purpose of the lessons, which was to promote reasoning. In all, 68 lessons were subsequently enacted in classrooms, and the research team visited and observed all these lessons. The main finding from the research was that shortly after the start of each lesson, more than two-thirds of the lessons quickly "declined" (the researchers' term) into lessons that simply involved students carrying out routine procedures (as a result of the teacher telling them what to do), working in ways that were non-systematic or even, in some cases, students engaging in non-mathematical activity (for example, coloring in diagrams).

Looking in more detail at the third of lessons where mathematical reasoning was maintained, the researchers identified a number of factors common across these lessons, factors that included building on students' prior knowledge and providing an appropriate amount of time (too much being as unproductive as too little). One notable factor for maintaining a focus on reasoning was evidence of "sustained pressure for explanation on meaning." Now teachers often, with good intent, want to make learning as pleasurable as possible, so "sustained pressure" sounds like the antithesis of this, but the work of the Nobel Prize winner, Daniel Kahneman, points to why such pressure might be necessary.

Kahneman proposes a model of our thinking as being either "fast" or "slow" [11]. Fast thinking is that which we do without having to reflect on it—knowing that seven times eight is fifty-six or that tan is sine divided by cosine. Slow thinking is more deliberate; it is the kind of thinking we have to engage in when doing mathematical reasoning. Over a number of studies, Kahneman has shown that moving from fast to slow thinking is often accompanied by a slight, but noticeable feeling of depression. The significance of this to teaching is that teachers recognize that moment when having given a class a challenging task, they can feel the energy in the room changing, and not in a positive direction. It seems reasonable to assume that one result of many of the lessons in Henningsen and Stein's research declining into routine procedures was a result of the negative energy arising from students moving into "slow" thinking with some teachers "easing" the classroom climate by pointing out what to do, while those teachers, applying "sustained pressure," helped students to work through their resistance.

A second possible reason for limited attention to mathematical reasoning in lessons may arise from what teachers attend to in planning. The researcher Ference Marton argues that in any teaching and learning encounter, there are "objects" of learning coming into being, and that these can either be direct objects or indirect objects [12]. Direct objects of learning are those aspects of lessons to which students are immediately attending. They are what students might say in response to the question "What did you do in math today?"— "We worked on subtraction," "We solved simultaneous equations," and so on. Marton argues that, whether intentional or not, every direct object of learning invokes a number of indirect objects of learning. For example, the student working through a page of simultaneous equations but without much sense of where such equations come from, or what the results mean, may come to learn that mathematics comprises a number of procedures that simply have to be committed to memory. That learning may not have been the direct intention of the teacher, but is an indirect consequence of the direct object of learning worked on.

In my experience, teachers, when planning mathematics lessons, often focus mainly on the direct object of learning—fractions, money, linear graphs and so on. Tasks are then collected together that address the direct object. While that might be the start of planning, it is also important to ask what the indirect outcomes might be of working on those tasks—what mathematical activity students are going to be engaging in as a result of working on the tasks. This is important in promoting mathematical reasoning, as reasoning cannot be taught directly. The direct task given to students is only a starting point. Having chosen a task, we have to think our way into what students are likely to do as a result of engaging with it, that is, what indirect learning might come about through working on the direct object. I recently visited Japan to observe not only a number of Lesson Study meetings (a form of professional development for a teacher where they observe and critique a carefully crafted lesson) but also a number of regular lessons. In the discussion following each of the lessons, a common focus was on what had been anticipated that the students might do mathematically, and whether or not this came about in the course of the lesson. Anticipating and working with the mathematical reasoning of the students was treated as more important than thinking about what the teacher had to do in the lesson.

Reasoning as a Habit of Mind

Another possible reason for why reasoning may get less attention in lessons is the perception that it has to be the focus of an entire lesson, and given the amount of "content" to cover, this may only be able to happen occasionally. In [13], Cuoco, Goldenburg, and Mark describe mathematical power as a "habit of mind," so if we want students to reason mathematically, then this needs to become a habit of mind, and, like any habit, is best developed little and often. Rather than the occasional "inquiry" lesson, reasoning chains—a series of linked, short activities—can engender such a habit of mind.

We looked at such a reasoning chain earlier in discussing whether or not 326 × 18 = 326 × 17 + 326 was true: the two simpler but structurally identical examples preceding this would move the conversation with students away from thinking about calculating answers to discussing and reasoning about the underlying structure. That example was structured around preserving the underlying structure while making the numbers involved appear to be more challenging. Another approach is to present a series, a chain, of calculations, where subsequent answers can be reasoned about based on the previous example. For instance:

$$160 \div 16$$
$$320 \div 16$$
$$320 \div 32$$

The reasoning here could go along the lines of, given that the answer to the first calculation is easy to calculate mentally, then if the dividend is doubled but the divisor unchanged, then the answer is going to double. The third example provides an opportunity to discuss why doubling both the dividend and divisor leaves the answer unchanged, and exploring why this is not the case with multiplication. Cathy Fosnot has written about such chains, and I recommend her work if you are interested to read more (*see*, for example, [14]).

Conclusion

If reasoning is going to become more central to most mathematics teaching and learning, then there are some shifts that need to come about in lessons. The most important shift is to move away from thinking that getting answers to problems is the end goal of the lesson. Answers have to be seen not as the end product of a lesson, but as the beginning, as an opportunity to examine the underlying mathematical generality. That involves a shift from asking:

How do I teach students to answer this problem?

to

What mathematical reasoning do I expect them to engage in as a result of working on this problem?

Currently the business of education with its focus on test results may not be conducive to such shifts, but we should not lose sight of the broader aims of education, aims that were important for John Dewey, more than 80 years ago, and, in our Internet age, may be even more important now:

While it is not the business of education . . . to teach every possible item of information, it is its business to cultivate deep-seated and effective habits of discriminating tested beliefs from mere assertions, guesses, and opinions. [15]

References

1. J. Kilpatrick, J. Swafford, and B. Findell, Eds. *Adding it up: Helping children learn mathematics*, Washington, DC: National Academy Press (2001).

2. Department for Education, *The national curriculum in England: Key stages 1 and 2 framework document*, London: Department for Education (2013).

3. A. N. Whitehead, *The aims of education and other essays*, New York: The Free Press (1929).

4. T. Nunes, P. Bryant, K. Sylva, and R. Barros, *Development of maths capabilities and confidence in primary school* [Research report], London: Department for Children, Schools and Families (DCSF) (2009).

5. T. Nunes, P. Bryant, R. Barros, and K. Sylva, The relative importance of two different mathematical abilities to mathematical achievement. *British Journal of Educational Psychology*, 82(1) (2012) pp. 136–156. https://doi.org/10.1111/j.2044–8279.2011.02033.x.

6. A. Arcavi, Symbol sense: Informal sense-making in formal mathematics. *For the Learning of Mathematics*, 14(3) (1994) pp. 24–35.

7. L. English and B. Sriraman, Problem Solving for the 21st Century. In *Theories of Mathematics Education*, B. Sriraman and L. English, Eds., (2010) pp. 263–290. Retrieved from http://link.springer.com/chapter/10.1007/978–3–642–00742–2_27.

8. T. Carpenter, E. Fennema, M. L. Franke, L. Levi, and S. B. Empson, *Children's Mathematics: Cognitively Guided Instruction*, Portsmouth, NH: Heinemann (1999).

9. S. Lamon, *Teaching Fractions and Ratios for Understanding*, 3rd ed., New York: Routledge (2011).

10. M. Henningsen and M. K. Stein, Mathematical tasks and student cognition: Classroom-based factors that support and inhibit high-level mathematical thinking and reasoning.

Journal for Research in Mathematics Education **28**(5) (1997) p. 524. https://doi.org/10.2307/749690.

11. D. Kahneman, *Thinking, fast and slow*, London: Allen Lane (2011).

12. F. Marton and S. Booth, *Learning and awareness*, Mahwah, NJ: Lawrence Erlbaum Associates (1997).

13. A. Cuoco, P. Goldenburg, and J. Mark, Habits of Mind: An Organizing Principle for Mathematics Curricula. *Journal of Mathematical Behavior* **15** (1996) pp. 375–402.

14. C. T. Fosnot and M. Dolk, *Young Mathematicians at Work: Constructing Fractions, Decimals and Percents*, Portsmouth, NH: Heinemann (2002).

15. J. Dewey, *How we think: A restatement of the relation of reflective thinking to the educative process*, New York: D. C. Heath and Company (1933).

Knowing and Teaching Elementary Mathematics—How Are We Doing?

ROGER HOWE

Knowing and Teaching Elementary Mathematics (KTEM) by Liping Ma [10], based on a comparative study of the mathematical understanding of U.S. and Chinese elementary mathematics teachers, was published in 1999. It landed in the U.S. mathematics education community like a bombshell. It also attracted substantial interest internationally—it has been translated into Chinese, Korean, Spanish, and Czech.

For much of the second half of the twentieth century, U.S. mathematics educators had been debating the role of the mathematical knowledge of teachers in fostering mathematics learning. Some studies [5] had indicated that the correlation between the amount of college mathematics training of teachers and the achievement of their students was very weak, and perhaps even negative. In [19], Lee Shulman proposed that perhaps content knowledge of teachers needed to be thought about differently, and that there is content knowledge that is specific to teaching, different in nature from the knowledge that engineers, scientists, and mathematicians need to ply their crafts. Taking a cue from Shulman, Deborah Ball [1–3] started investigating what might be the nature of that knowledge, which she later [4] called Mathematical Knowledge for Teaching (MKT). With some colleagues at the National Center for Research on Teacher Learning (NCRTL) at Michigan State University, she developed a set of interview questions for gauging knowledge that plausibly could contribute to MKT and which could only be correctly answered with a good understanding of several key ideas of elementary math: the reasons behind the computational algorithms for subtraction and multiplication, the meaning of division by fractions, and the relationship between area and perimeter [9].

Liping Ma took these four questions, translated them into Chinese, and went to China to pose them to a group of Chinese teachers. The teachers were not randomly selected, but did represent several types of teaching situations: for example, different grade levels, urban and rural. She compared the results of her interviews to those of a group of U.S. teachers who had been interviewed with the NCRTL questions. These teachers had been described as "better than average" in the United States by an expert who had worked with them.

The results from the two sets of interviews were strikingly different: The Chinese teachers provided dramatically superior responses, especially to the more challenging questions. The group of Chinese teachers collectively showed a level of expertise in elementary mathematics that was off the charts relative to U.S. experience.

Dramatic as these results were, KTEM went considerably beyond a simple report of comparative performance. Ma formulated a concept of "profound understanding of fundamental mathematics" (PUFM) that captured the knowledge level of the best of the Chinese teachers, and she investigated how these teachers arrived at this level. This further work made clear that, in some sense, it was not fair to compare American teachers to their Chinese counterparts. Although the Chinese teachers benefited from good training in K-12 and college, their approach to and achievement of PUFM involved continued learning over their years of teaching, and this learning was enabled by their work conditions, conditions that were drastically different from those of U.S. teachers, especially with respect to supporting professional growth.

For one thing, even at the elementary level, most of the Chinese teachers were specialists who were only responsible for learning mathematics and could concentrate on it. For another, many of the Chinese teachers had "taught in rounds," following the same class through several years, so that they had a perspective on how mathematics learning develops over a range of years. Third, at the expense of running larger classes (more like 40-plus than 30-minus), the Chinese teachers were required to be in the classroom only half or less of each school day. The rest of the day they could spend grading, working with students who needed extra attention, studying the teaching materials (mainly textbooks, officially produced and highly focused on the mathematical issues of each lesson, and also the Chinese mathematics standards). Most saliently, they had time and opportunity to talk with colleagues about teaching issues. These discussions were facilitated by the fact that usually the math teachers had a common office where they would go to work when not teaching.

Thus, the Chinese elementary mathematics teachers enjoyed a workday that promoted continual learning, and over the years, this helped many of them to achieve an impressive grasp of the key issues of their elementary mathematics curriculum. At the time that KTEM was published, it was not possible to evaluate directly how this superior teacher knowledge translated into student achievement (although Ma did present evidence that it did). However, when Shanghai was ranked number 1 in the world when it first (2012) participated in the Program for International Student Assessment (PISA) [15, p. 46], the relationship was substantiated in a more official way.

At about the time of publication of KTEM, the National Academies of Science was asked to review and summarize the research on mathematics education and to make recommendations for improving it. The report resulting from the study was *Adding It Up*, published in 2001 [13]. Two (of the five) recommendations in the report recognized the need for systematic, long-term effort to become a proficient mathematics teacher, and called for structures to address this need:

- Teachers' professional development should be high quality, sustained, and systematically designed to help all students develop mathematical proficiency. Schools should support, as a central part of teachers' work, engagement in sustained efforts to improve their mathematics instruction. This support requires the provision of time and resources.
- The coordination of curriculum, instructional materials, assessment, instruction, professional development, and school organization around the development of mathematical proficiency should drive school improvement efforts.

Somewhat later, National Research Council work also showed that there are even more systematic structures in place in the Chinese school system to support teacher learning. In 2009, ten years after publication of KTEM, the U.S. National Commission on Mathematics Instruction (USNC/MI) held a joint U.S.-China workshop exploring the career structure of mathematics teachers in the two countries [14]. This workshop revealed a substantial infrastructure supporting teacher learning in China. In general, a Chinese teacher's career is more structured and offers more opportunities for advancement than is the case in the United States. A standard teaching career moves through three ranks: second rank teacher, first rank teacher, and senior or master teacher. The movement from one rank to the next involves various formal requirements, including a substantial amount of continuing education, and for the master rank, publication of research, often on effective methods for teaching specific topics. Novice teachers are assigned senior mentors, who give advice on lesson plans and attend and give feedback on lessons taught by the novice teachers, who in turn visit the classroom of their mentor teacher to observe and absorb effective teaching moves.

Furthermore, beyond the normal three ranks, there are opportunities to become a "super-rank" teacher. The rules governing these higher ranks are not uniform across China, but in a city like Shanghai (which has a population larger than New York State), the super-rank structure is quite systematic, and super-rank teachers form a specified (rather small, under 1%) portion of all teachers. To become a super-rank teacher, one must publish on good teaching techniques for selected topics and also compete in teaching contests. Besides teaching students, the job of super-rank teachers involves running professional development for their normal rank colleagues. In this way, knowledge of effective teaching techniques and responsibility for propagating them is embodied in the teaching corps itself.

Consistent with the existence of the master and super ranks for teachers, with responsibility for professional development, there is a substantial system of continual professional development in place. Teachers participate in research groups in their schools or districts, and many research groups work on new teacher induction, preparing for rising to master teacher level, and discussion of effective teaching of various specific topics. The strong support for helping teachers improve highlighted in the 2009 USNC/MI workshop is also documented in a World Bank report [21].

The United States has nothing comparable to this level of support for continued learning. Teachers who are determined to make the effort to improve are mainly left to their own devices. Moreover, there are few mechanisms for sharing anything that is learned.

One might have hoped that these structural deficits in the organization of U.S. mathematics education, pointed to so clearly by Liping

Ma and further reinforced by *Adding It Up* and then again by USNC/ MI, would have gotten the attention of U.S. educational policymakers and resulted in widespread reforms. Unfortunately for our teachers and their students, this has not happened. For example, when the Common Core State Standards for Mathematics (CCSSM) program was rolled out around 2010, it was formally adopted by most states, but little to nothing was done to enable teachers to teach in the ways advocated by the CCSSM. New York did commission a project to create a text consistent with CCSSM, which resulted in Eureka Math (aka Engage New York) [8]. This has a lot of good points, including teacher editions with extensive discussions of the relevant mathematical ideas; but these teacher editions mostly run between 1,000 and 2,000 pages per grade. They clearly require intensive study, but arrangements to promote that study are mostly lacking.

The lack of support for teacher improvement in U.S. education is more or less unchanged in the two decades since Liping Ma showed us the dramatic deficits of teacher mathematics expertise that results from it, and international examinations such as Trends in International Mathematics and Science Study (TIMSS) and the Programme for International Student Assessment (PISA) have shown us that this translates into deficits in student learning. U.S. education policymakers have not learned from this cross-cultural research; they have preferred a top-down scheme of standardized tests, with punishments for submandated performance. So it should come as no surprise that, as recorded in the latest PISA (2018) results [18], our children continue not to learn mathematics very well.

Some might argue that these kinds of deficits are culturally determined, so that there is no possibility of matching Chinese levels of performance. However, there are examples closer to home of school systems that achieve better results in mathematics. In particular, Canada has consistently done better in international comparisons than the United States [17]. Even more relevant for us is the experience of Massachusetts, which in the early 1990s had mathematics achievement that was fairly typical of the United States, but which decided that that was not good enough, and embarked on a long-term effort to improve. The Massachusetts Education Reform Act of 1993 provided for increased funding to make opportunities to learn more equitable, substantial professional development, and higher mathematics requirements for teacher certification (thus impacting teacher preparation programs) [6]. It worked. Massachusetts now, considered as a separate country, has test scores significantly above the U.S. average [20], [16], [11], [7]. Massachusetts also comes out at the top in the National Assessment of Educational Progress (NAEP), the United States' own instrument for assessing student achievement. For example, in the 2019 edition of NAEP, Massachusetts eighth graders had the highest overall average of any state, and perhaps more significantly, 48% of them scored at the proficient or advanced level [12]. Only two other states had over 40% at these levels. The rest of the country should take some lessons.

Acknowledgments

The author is grateful to the referee for a careful reading and many helpful suggestions, which substantially improved the paper. Thanks also to Richard Bisk, Albert Cuoco, and Thomas Fortmann for information about Massachusetts.

References

[1] D. Ball, *The Subject Matter Preparation of Prospective Teachers: Challenging the Myths*, National Center for Research on Teacher Learning, East Lansing, MI, 1988.

[2] D. Ball, *Teaching Mathematics for Understanding: What Do Teachers Need to Know about the Subject Matter?*, National Center for Research on Teacher Learning, East Lansing, MI, 1989.

[3] D. Ball, Prospective elementary and secondary teachers' understanding of division, *J. Res. Math. Ed.* **21** (1990), 132–144.

[4] D. L. Ball, S. T. Lubienski, and D. S. Mewborn, Research on teaching mathematics: The unsolved problem of teachers' mathematical knowledge. In V. Richardson (Ed.), *Handbook of research on teaching* (4th ed.), Macmillan, New York, 2001, pp. 433–456.

[5] *Critical variables in mathematics education: Findings from a survey of empirical literature*, Mathematical Association of America and National Council of Teachers of Mathematics, Washington, DC, 1979.

[6] M. Chester, *Building on 20 Years of Massachusetts Education Reform*, Massachusetts Board of Education, Boston, 2014.

[7] *Bringing it back home* (Briefing paper #410). Available at https://www.epi.org/publication/bringing-it-back-home-why-state-comparisons-are-more-useful-than-international-comparisons-for-improving-u-s-education-policy/.

[8] Eureka Math. Available at https://greatminds.org/math.

[9] A Study Package for Examining and Tracking Changes in Teachers' Knowledge. Available at https://eric.ed.gov/?id=ED359170.

[10] Liping Ma, *Knowing and Teaching Elementary Mathematics*, Lawrence Erlbaum Assoc. Inc., Mahwah, NJ; London, 1999.

[11] *Massachusetts students score among world leaders on PISA reading, science and math tests* [press release]. Available at www.doe.mass.edu/news/news.aspx?id=24050.

[12] The Nation's Report Card. Available at https://www.nationsreportcard.gov/ndecore/xplore/NDE.

[13] J. Kilpatrick, J. Swafford, and B. Findell (Eds.), *Adding It Up: Helping Children Learn Mathematics*, Mathematics Learning Study Committee, Center for Education, Division of Behavioral Sciences and Education, National Academies Press, Washington, DC, 2001.

[14] A. Ferreras and S. Olson, rapporteurs, A. E. Stein, Ed., *The Teacher Development Continuum in the United States and China, Summary of a Workshop*, National Academies Press, Washington, DC, 2010.

[15] *PISA 2012 Results: What Students Know and Can Do: Student Performance in Mathematics, Reading and Science* (Volume I, Revised edition, February 2014), PISA, OECD Publishing. dx.doi.org/10.1787/9789264208780-en.

[16] PISA Massachusetts results, 2015, Available at https://www.oecd.org/pisa/PISA-2015-United-States-MA.pdf.

[17] PISA Canada results, 2018, Available at https://www.oecd.org/pisa/publications/PISA2018_CN_CAN.pdf.

[18] PISA United States results, 2018, Available at https://www.oecd.org/pisa/publications/PISA2018_CN_USA.pdf.

[19] L. Shulman, Those who understand: Knowledge growth in teaching, *Educational Researcher* **15** (1986), 4–14.

[20] TIMMSS results for Massachusetts, 2011, Available at https://nces.ed.gov/timss/pdf/results11_Massachusetts_Math.pdf.

[21] *World Bank Study Shows Shanghai's #1 Global Ranking in Reading, Math, & Science Rests on Strong Education System with Great Teachers* [press release]. Available at https://www.worldbank.org/en/news/press-release/2016/05/16/world-bank-study-shows-shanghais-1-global-ranking-in-reading-math-science-rests-on-strong-education-system-with-great-teachers.

Tips for Undergraduate
Research Supervisors

STEPHAN RAMON GARCIA

I was recently asked to contribute a paper to a forthcoming *Foundations for Undergraduate Research in Mathematics* (FURM) volume [1]. The article contained, in addition to several lengthy case studies, a list of brief recommendations for undergraduate research supervisors. Several colleagues suggested developing these general principles into a separate, self-contained article. What follows is an expanded version of the recommendations introduced in [1]. I focus here on how a supervisor of undergraduate research can direct students to fruitful research projects and shepherd the subsequent results through to publication.

This is not an exhaustive guide, simply a list of recommendations based upon my individual experience. It is not a scholarly endeavor, but rather a personal impression of tips, tricks, ideas, and perspectives that I have found useful in developing a sustainable undergraduate research pipeline. However, these recommendations are not necessarily universal. I am a pure mathematician whose experience supervising undergraduate research stems mostly from academic-year projects, often as part of a senior exercise. Thus, this advice is not precisely tailored for those running research experiences for undergraduates (REUs). Moreover, I have generally focused on producing published research articles with my students. This may not correlate exactly with the reader's goals and expectations. Nevertheless, I hope that the reader will find at least some of the following reflections useful when working with their own students.

1. **Time is a luxury you don't have.** There are major differences between the research undertaken by a graduate student in pursuit of a Ph.D. and an undergraduate student working

on a senior thesis or summer research project. The Ph.D. student takes graduate-level courses and trains for several years in a specific area under the supervision of the advisor, who is a leading expert in the field. In contrast, the undergraduate may have little or no coursework related to the topic, especially if the subject matter falls outside of the standard curriculum. Think carefully whether your preferred topic is realistic for a student to engage with in the given time frame. The research supervisor must be nimble, able to adjust to circumstances and constraints. Finding projects that are accessible and realistic for your students takes time and thought.

2. **You are not an old dog.** You can learn new tricks. Undergraduate research might require you to take a plunge into unfamiliar territory. Learning new mathematics with students is one of the great personal benefits of undergraduate research. Use it as an opportunity to get out of your Ph.D.-thesis shell. The further down a narrow research avenue you travel, the more difficult, specialized, and potentially inaccessible the problems become. To gain the flexibility and perspective necessary to develop viable undergraduate research projects, you need to expose yourself to new ideas. Do not be afraid to wade into uncharted waters for a project. Embrace the challenge!

3. **You know more than they do.** Even the cleverest of your undergraduate students has a perspective that is largely limited by standard coursework. Students who have taken part in mathematical enrichment programs may have a larger repertoire of tricks and tools, but they are not up to date with current research. On the other hand, you have an advanced degree in the mathematical sciences and are trained to conduct independent research. You are perfectly qualified to lead students on an exciting mathematical adventure. With your training and experience, you can stay a few steps ahead of your students, even in unfamiliar territory. In fact, your ideas and greater perspective may be useful, even in an unrelated field. Trust yourself!

4. **Nobody knows everything.** You do not have to be a world expert on a topic to get your students started on a project in

that area. Even in your own field of expertise, you probably cannot answer every question that comes your way. When learning new mathematics, there are bound to be points of confusion, unclear definitions, and seeming contradictions. Working through ambiguity and uncertainty is a key trait of a good mathematician. Seeing a mathematician learning new material provides your students a valuable lesson.

5. **You can be human.** At some point, your students may discover that you are human. Even truly great mathematicians encounter problems that stump them, so there is no shame in admitting that you are stuck. Students have a tendency to doubt themselves when they hit a snag. They may have unrealistic impressions of how "real" mathematicians operate and judge themselves against an unattainable ideal. Seeing a professor deal with pitfalls and roadblocks is an excellent learning experience for students.

6. **Follow their passions.** A student can fall in love with a particular topic and become obsessed. A theorem or example can pique their interest and spark their imagination. Students have more energy and enthusiasm for projects they love than for projects they are simply assigned by decree. It can sometimes be a good idea to permit students to follow their passions, as long as the journey is tempered by realism. Instead of focusing on a major open problem, you might steer students to variants or closely related topics that still hold their interest. A student who is keenly interested in a subject, paper, or theorem can provide you with an important opportunity for exploration. Learning something new together can be a great experience. Your students might lead you into terrain that you never imagined exploring.

7. **Search for fertile ground.** Topics and questions that have been combed over by experts for decades are like dry, parched earth. With work, they might still yield fruit, although the field will be tough, perhaps too tough for an undergraduate to obtain results during the course of a summer or a senior thesis. You do not want your students competing head-to-head against leading experts who have dedicated their lives to a subject. Instead, you need to find different angles and

new questions that are adjacent or parallel to where the "big names" are working. You need to be close enough to established work that your project is respectable and cannot be easily dismissed, but not so close that your students will be constantly scooped by more experienced mathematicians.

8. **Andrew Wiles' approach is not for everyone.** Sir Andrew Wiles spent seven years in near isolation working on Fermat's Last Theorem before he completed his proof (and even then, it initially contained a gap). Most mathematicians, myself included, are not capable of banging their heads against a difficult problem, day in and day out, for long periods of time before frustration and boredom set in. These feelings are nothing new for the professional mathematician. However, they can rapidly erode students' confidence and drive them to question whether they belong in mathematics at all. Students should not be given the impression that the Wiles-style approach is how most mathematics is done.

9. **Pivot when returns diminish.** Our students lack the experience and instinct of the professional mathematician, who has written a dissertation over the course of several years and many sleepless nights. They cannot always tell which problems are too difficult and which are realistic targets. Be prepared to pivot and shift to more feasible problems if necessary. Always make sure that your students are engaged. When mathematics stops being fun and exciting, it is time to change things up. There should be other options on the table, other questions and variations on the original theme that might prove more tractable.

10. **Focus more broadly.**[1] An intimidating and often unhelpful way to introduce students to a topic is to hand them a paper and say "read this thoroughly, understand it, and report back to me next week." One does not learn mathematics by reading—one learns by doing. Students need to hit the ground running. Have your students *skim* through a few papers in the area. Are there gentle expository papers on the topic? Those are ideal.

11. **Play 20 questions.** Ask your students to devise a list of 20 questions inspired by the reading. They come to a topic fresh

and are able to ask the naïve questions that we long ago learned to suppress. Sometimes these questions can be the starting point for an entire new line of research. Encourage your students to find interesting variants of existing questions. They need not be directly related to the topic; tangents should be especially appreciated. Even if only 10% of the questions that come up pan out, the exercise will have been a smashing success. Moreover, the students will feel ownership and pride in a project they develop themselves.

12. **Everything is negotiable.** Students should understand that every aspect of a question or problem is fluid and changeable. The original problem might prove too difficult. Perhaps there is a change of context that renders a tractable problem. What happens if some of the hypotheses of your conjecture are changed? What about looking at the same question in a different setting? Be open to sudden changes in direction and embrace the opportunity to pursue promising leads. The more questions that you and your group generate, the larger your interface with the unknown is. Instead of chipping away relentlessly at one specific problem, your students can investigate multiple points of entry, in the hope that one of them will give way. You need only gain traction on a single problem before new results start pouring in.

13. **Turn lemons into lemonade.** Sometimes things do not work out as planned. Your hunches may prove incorrect, and your conjectures might be false. Do not think of this as a failure, but rather as an unusual flavor of opportunity. You must show your students how to turn things around and salvage something from the apparent disaster. Think of it as a challenge: you need to turn the negative into a positive, a failure into success. Perhaps a counterexample is more interesting than the conjecture itself. Perhaps the proof broke down in an interesting way and the counterexample points toward the correct result. Can you tighten the conjecture to make it correct? Can you count, predict, or characterize the counterexamples? There are many directions one can pivot to after an initial disappointment if one does not view a setback

as a final roadblock, but rather as a detour down an unfamiliar road.

14. **Complement your research.** "Undergraduate research" and "research" are not necessarily distinct things. There may be parts of your own research that could benefit from student involvement. Perhaps there is a tedious example that you have not had the patience to work through, even though you suspect that it will be straightforward. Maybe you have a lot of numerical evidence to gather, but have not had time to write the program and run the code. Perhaps you feel that a certain result is true and that you could prove it if only you had a few hours to grind through the details. These are perfect tasks for student researchers. What might take you hours could take a student weeks, to the benefit of both parties. What you find tedious, a student might find new and enlightening. Your personal research problems might have simpler versions suitable for students. Conversely, student research might inspire and inform your own work. It might suggest questions that you have not considered before or require you to learn new material that opens up novel avenues in your work.

15. **The computer is your friend.** Mathematical software is an excellent tool for undergraduate research. Software can help you visualize data and see patterns that would otherwise escape human notice. Students can jump right into experimentation and conjecture making, even before they fully understand the theoretical underpinnings of the project. This is a lot more fun than the traditional "read this paper and report back on what you learned" approach. Students catch on quickly to technology, especially these days when many mathematics majors take at least an introductory computer science course. They can often learn what they need to know on the fly, since there is no shortage of how-to websites and instructional videos for Mathematica, MATLAB, Maple, Sage, and so forth. Moreover, websites like Stack Exchange answer many rookie questions and often provide snippets of code for common tasks. Software lets your students hit the ground running.

16. **Build upon previous success.** If an earlier project worked out, you might as well continue it. Are there variations, generalizations, or extensions worth pursuing? The roadmap provided by a previous project can suggest a series of discrete tasks for the next generation of students. However, even apparently straightforward generalizations may lead to unexpected results. It almost always pays to investigate leads, even if they look simple and unassuming. Often there are additional complications that turn up once one works through the details. Variations are often less transparent and more difficult than one suspects. Perhaps the obvious generalization fails because of a subtle counterexample? Complications make for more interesting follow-up projects.

17. **Feel free to hand wave.** Proofs are not written on the first attempt. Rather, they are the result of many abortive tries, lots of scratch paper, and much hand wringing. However, students do not always see the labor and long hours that go into a result. Standard coursework suggests that each theorem follows from the previous one, in logical progression and in bite-sized chunks. How do we professors establish our research results? We usually have a general idea, informed by experience and intuition. We do not typically see all of the details at the beginning, but have only a rough idea of an approach. Students need to learn how we go about proving things. Heuristic arguments, informal reasoning, and numerical evidence can get a project started. The skeleton of an argument can be fleshed out later. A hand-wavy argument can help to parcel out different pieces of the project to multiple students. Perhaps a student could prove an important subcase or a crucial lemma, or develop an instructive example.

18. **You do not have to go it alone!** Students should learn early on that mathematics is a social endeavor. Contrary to popular opinion, most mathematicians do not pursue difficult problems in total isolation, emerging only years later to declare their triumph. We communicate through networks of collaborators and online message boards like MathOverflow, and we attend conferences, seminars, and colloquia. We are not solitary animals, although students do not always realize

it. Most mathematicians love to talk about their work. In fact, they are often grateful for any attention their research generates, and they will typically respond to a sincere inquiry about one of their papers. Feel free to write to a mathematician out of the blue if their work is relevant to the project. There is much to gain, and nothing to lose. In the worst-case scenario, you have wasted a few minutes of your time. In the best-case scenario, you might get a crucial insight into your problem, understand a confusing point from a paper, or even gain a new collaborator.

19. **Know your audience.** There are many suitable outlets for undergraduate research. Although *Involve* is probably the most well-known example, there are many other options, such as the *Rose-Hulman Undergraduate Mathematics Journal, SIAM Undergraduate Research Online*, the *Pi Mu Epsilon Journal*, the *PUMP Journal of Undergraduate Research*, and the *Minnesota Journal of Undergraduate Mathematics*. On the other hand, getting undergraduate research published in mainstream research journals requires some strategy. One should have an idea of the audience before completing the final writeup since the introduction and style of the paper will depend upon the target journal. Hopefully, your project involves some relatively mainstream ideas and buzzwords, even if the work is not at the bustling center of the field. You must make some effort to relate your work to the existing literature, hopefully to papers published in the sort of journals you are considering. If your paper is more of an isolated curiosity or an oddball result, the chances of acceptance are lower.

20. **Use the modularity principle.** Not every student project is publishable. With care and foresight, however, many student projects can contribute meaningfully to a publication. For example, a student might have worked out the details of an instructive example, amassed a significant amount of computational evidence, or obtained a small improvement on a known result. None of these might be publishable in and of themselves. However, all three could form a "module" in a larger work. A substantial article can be built up over the course of several undergraduate research projects. It may take

a few "generations" of students before enough compelling material is collected to form a publishable research paper. You might have to hold on to a clever example, minor theorem, or partial result for a while. Keep in touch with former students in case you need to contact them later. Most of them will be delighted to learn that their long-forgotten senior thesis project will form a crucial part of a research paper.

21. **Do not drag your feet.** Time is of the essence for undergraduate research students. Some of them might be applying to graduate school. A good publication can enhance an otherwise lackluster graduate school application. Other students may be looking for employment in the "real world" soon, and a math paper might just spice up their resumés. If you have something publishable, do not sit on it! For a tenured professor, there is little consequence in waiting a month or two to get around to writing something up. However, students (and untenured professors) are on a different clock. It is your duty to help complete the writeup and, if the results warrant it, submit the paper to a journal in a timely fashion. You should be the one responsible for posting it on the arXiv, ultimately deciding upon a journal, and submitting it. Only you can prevent your own foot dragging. Your students did a lot of hard work. You owe it to them to see things through in a timely manner.

22. **Is there an opportunity for exposition?** The literature on a particular topic develops organically. Important results might be strewn throughout dozens of articles spread over several decades. In many cases, there is not a standard reference to which you can refer your students. This is an opportunity to write an expository article, or a long survey article, on the subject. You and your students will probably conduct a literature review early in the project and take copious notes about the fundamentals anyway. It makes sense for this background work to pay off. Proper exposition may require detailed examples, fleshed-out proofs, extensive computations, and so on. Each of these pieces can provide a small project for a student. But where could such an article be published? In pure mathematics, conference proceedings or book chapters are common options. One might consider the journal *Expositiones*

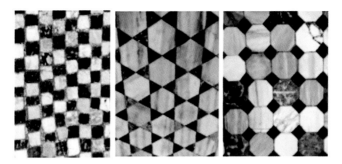

FIGURE 1 from "Cosmatesque Design and Complex Analysis" (Pomerantz)

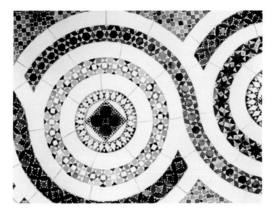

FIGURE 2 from "Cosmatesque Design and Complex Analysis" (Pomerantz)

FIGURE 3 from "Cosmatesque Design and Complex Analysis" (Pomerantz)

FIGURE 4 from "Cosmatesque Design and Complex Analysis" (Pomerantz)

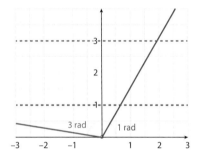

FIGURE 7 from "Cosmatesque Design and Complex Analysis" (Pomerantz)

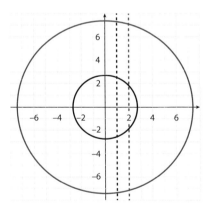

FIGURE 8 from "Cosmatesque Design and Complex Analysis" (Pomerantz)

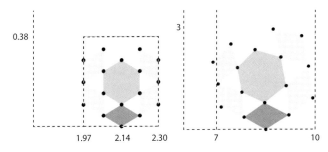

FIGURE 9 from "Cosmatesque Design and Complex Analysis" (Pomerantz)

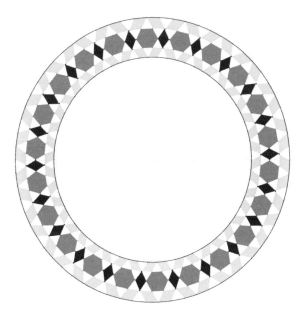

FIGURE 10 from "Cosmatesque Design and Complex Analysis" (Pomerantz)

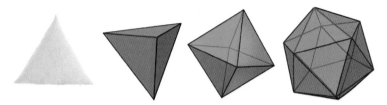

FIGURE 1 from "Hyperbolic Flowers" (Trnkova)

FIGURE 3 from "Hyperbolic Flowers" (Trnkova)

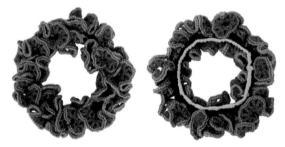

FIGURE 7 from "Hyperbolic Flowers" (Trnkova)

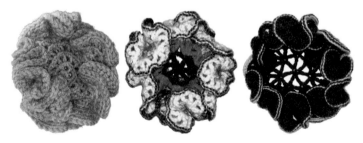

FIGURE 10 from "Hyperbolic Flowers" (Trnkova)

Figure 1 from "Modeling Dynamical Systems for 3D Printing" (Lucas/Sander/Taalman)

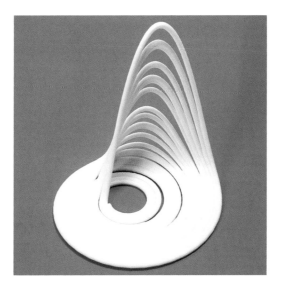

Figure 2 from "Modeling Dynamical Systems for 3D Printing" (Lucas/Sander/Taalman). Photo credit: Edmund Harriss.

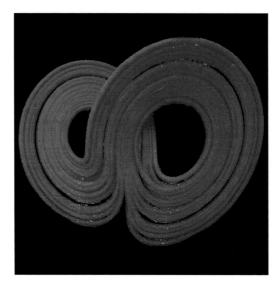

FIGURE 3 from "Modeling Dynamical Systems for 3D Printing" (Lucas/Sander/Taalman)

FIGURE 4 from "Modeling Dynamical Systems for 3D Printing" (Lucas/Sander/Taalman). Photo credit: Edmund Harriss.

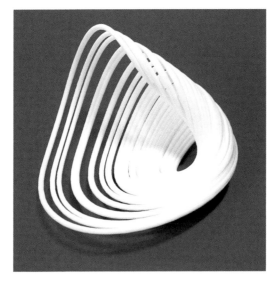

FIGURE 5 from "Modeling Dynamical Systems for 3D Printing" (Lucas/
Sander/Taalman). Photo credit: Edmund Harriss.

FIGURE 6 from "Modeling Dynamical Systems for 3D Printing" (Lucas/
Sander/Taalman)

Figure 7 from "Modeling Dynamical Systems for 3D Printing" (Lucas/Sander/Taalman)

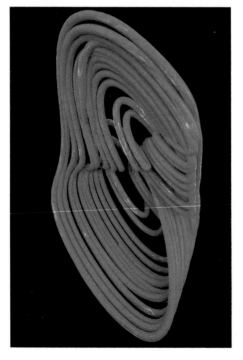

Figure 8 from "Modeling Dynamical Systems for 3D Printing" (Lucas/Sander/Taalman)

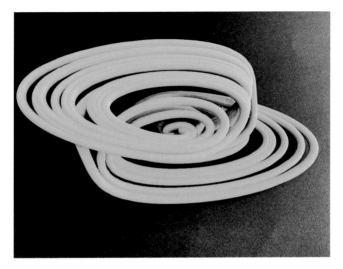

FIGURE 9 from "Modeling Dynamical Systems for 3D Printing" (Lucas/Sander/Taalman)

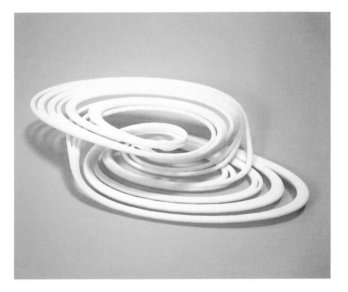

FIGURE 10 from "Modeling Dynamical Systems for 3D Printing" (Lucas/Sander/Taalman). Photo credit: Edmund Harriss.

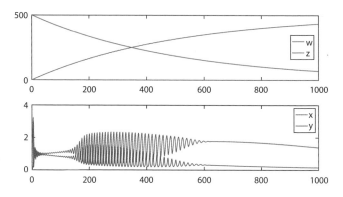

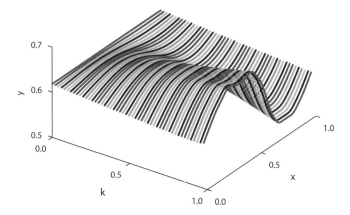

Figure 11 from "Modeling Dynamical Systems for 3D Printing" (Lucas/ Sander/Taalman)

Figure 18 from "Modeling Dynamical Systems for 3D Printing" (Lucas/ Sander/Taalman)

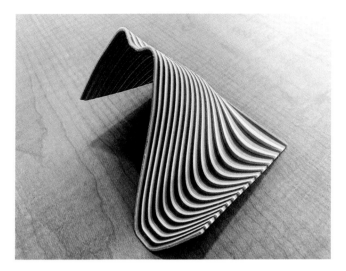

FIGURE 19 from "Modeling Dynamical Systems for 3D Printing" (Lucas/
Sander/Taalman)

FIGURE 21 from "Modeling Dynamical Systems for 3D Printing" (Lucas/
Sander/Taalman). Photo credit: Edmund Harriss.

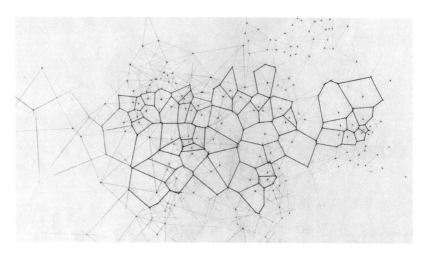

A Voronoi diagram. From "Scientists Uncover the Universal Geometry of Geology" (Sokol). From Fred Scharmen

The Geometry of Mars

To analyze a surface — in this case,
the honeycombed surface of a Martian crater
— researchers map out all edges and vertices.
They count the number of vertices per cell,
and how many cells share a given vertex.

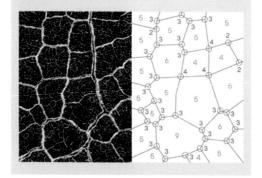

From "Scientists Uncover the Universal Geometry of Geology" (Sokol). Samuel Velasco / *Quanta Magazine*. Based on graphics from doi. org/ 10.1073/pnas.2001037117; Martian surface: NASA/JPL-Caltech/University of Arizona

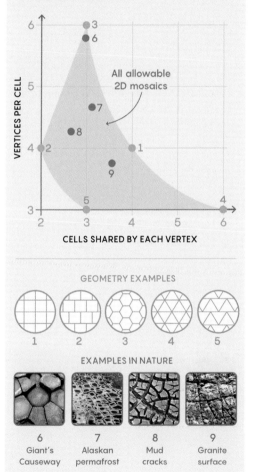

The Entire Tile Universe

What convex 2D shapes can fill a space without any overlaps or gaps? The possibilities can be mapped by counting the average number of vertices per shape and the number of shapes that share each vertex.

All allowable 2D mosaics

VERTICES PER CELL

CELLS SHARED BY EACH VERTEX

GEOMETRY EXAMPLES

1 2 3 4 5

EXAMPLES IN NATURE

6 Giant's Causeway
7 Alaskan permafrost
8 Mud cracks
9 Granite surface

FROM "Scientists Uncover the Universal Geometry of Geology" (Sokol). Samuel Velasco / *Quanta Magazine*. Based on graphics from doi.org/10.1073/pnas.2001037117; spot images: Lindy Buckley; Matthew L. Druckenmiller; Hannes Grobe; Courtesy of János Török

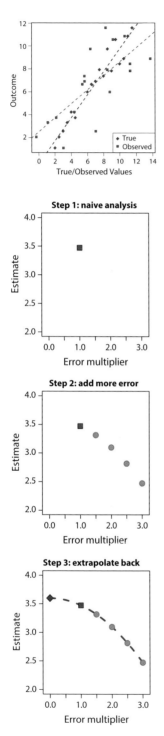

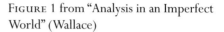

FIGURE 1 from "Analysis in an Imperfect World" (Wallace)

FIGURE 2 from "Analysis in an Imperfect World" (Wallace)

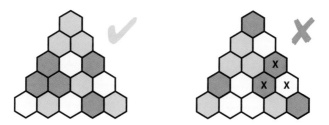

Figure 1 from "Tricolor Pyramids" (Siehler)

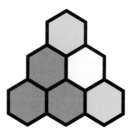

Figure 4 from "Tricolor Pyramids" (Siehler)

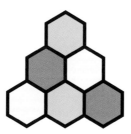

Figure 5 from "Tricolor Pyramids" (Siehler)

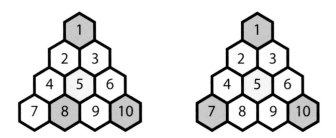

Figure 6 from "Tricolor Pyramids" (Siehler)

+	0	1	2
0	0	1	2
1	1	2	0
2	2	0	1

×	0	1	2
0	0	0	0
1	0	1	2
2	0	2	1

Table 1 from "Tricolor Pyramids" (Siehler)

Mathematicae. You can aim high and shoot for the *Notices of the American Mathematical Society* or the *American Mathematical Monthly*, although one should always have a fallback plan in case of rejection.[2]

23. **Reach out to new communities.** Projects can pull you in new directions. You might need to learn a bit of new mathematics in order to supervise your research group. Make an effort to meet people in the field. Are there seminars nearby? Attend conferences and meet a few new people. Most mathematicians are welcoming to newcomers in an area. After all, new blood guarantees the vitality and long-term viability of the subject. Ask to give a short talk in a special session at an American Mathematical Society conference. Ask to be put on the mailing list for annual conferences in your new subfield. Urge your students to give poster presentations at local or national meetings. You and your students need to become known, and this takes conscious effort.

24. **Get them recognized.** Suppose that you and your students have just completed a successful project. Perhaps it is a fine senior thesis, an excellent REU project, or a published research paper. Now comes the final, and possibly most important step, in promoting your student's research. Does your institution or department have a research prize for undergraduate students? If so, nominate them! It is up to you, since likely nobody else knows about the project or cares enough to nominate your student. Your institution probably has a large and well-funded communications staff that handles media relations, Twitter feeds, Facebook profiles, and so forth. If there is a compelling story to tell, your institution's communications staff wants to know. They are always on the lookout for good stories that highlight student research. This is part of their job! So get to know your communications staff. There will be a fancy web page that lists the contact information of the key communications personnel. Get to know them and clue them in to any exciting work done by your students.

The preceding list is not a foolproof recipe for successful undergraduate research mentoring. It is a reflection of my personal experiences and preferences. However, I hope that these general ideas have broader merit and that at least some of these points will apply to anyone in the mathematical sciences who is considering supervising an undergraduate research project.

Acknowledgments

Thanks go to Steven J. Miller and Mohamed Omar for helpful comments. This work was partially supported by NSF grant DMS-1800123.

Notes

1. An oxymoronic instruction from a reviewer who dismissed a colleague's grant proposal.
2. https://mathoverflow.net/questions/15366/which-journals-publish-expository-work discusses a few other options.

References

[1] Garcia, S. R. (2020) Lateral Movement in Undergraduate Research: Case Studies in Number Theory. In *A Project-Based Guide to Undergraduate Research in Mathematics*, Harris, P. E., Insko, E., and Wootton, A., Eds., Foundations for Undergraduate Research in Mathematics, pp 203–234. Birkhäuser, Cham, Switzerland. https://doi.org/10.1007/978-3-030-37853-0.

"The Infinite Is the Chasm
in Which Our Thoughts Are Lost":
Reflections on Sophie Germain's Essays

ADAM GLESSER, BOGDAN D. SUCEAVĂ,
AND MIHAELA B. VÂJIAC

Sophie Germain (1776–1831) is quite well known to the mathematical community for her contributions to number theory [17] and elasticity theory (e.g., *see* [2, 5]). On the other hand, there have been few attempts to understand Sophie Germain as an intellectual of her time, as an independent thinker outside of academia, and as a female mathematician in France facing the prejudice of the time of the First Empire and of the Bourbon Restoration, while pursuing her thoughts and interests and writing on them. Sophie Germain had to face a double challenge: the mathematical difficulty of the problems she approached and the sociocultural context of her time, which never fully supported her interests, never appropriately rewarded her, and never allowed her to enjoy the recognition she deserved. In our attempt to understand the innermost Sophie Germain, we also try to grasp the place of her personality within her time and historical period. We will argue that she represents a unique case in both the history of mathematics and the context of Western European intellectuals at the beginning of the nineteenth century, deserving a further exploratory study of the connections of her work with the ideas of her time.

Deservedly, toward the end of the twentieth century, Sophie Germain's works received attention in several thorough and useful inquiries, e.g., [12, 13, 15]. However, a specialist during this same period deemed her work as not worthy of glory [18], and she was even described as a "minor author" [19]. This is why we posit that further analysis and careful discussion of her intellectual achievements—mathematical or otherwise—is necessary. Our goal is to better assess her important contributions and to invite the consideration of her achievements and vision in the same manner as the ones of her contemporaries such as Gauss, Lagrange, Cauchy, and Poncelet.

We will start our argument with the uncontroversial fact that Sophie Germain is the mathematician who first introduced the concept of mean curvature [9]. This concept is a fundamental one in differential geometry [3], and its introduction generated a profound discussion about minimality in the geometry of submanifolds that is still relevant today and that led to the study of a plethora of new curvature invariants [3]. This turning point in differential geometry led to, among thousands of other results, the recent investigations on Willmore energy, which in turn brought us fundamental new results in differential geometry, such as the solution of the Wilmore Conjecture by Marques and Neves [14]. Germain's work also became a historical starting point for the 2018 Abel Prize winner Karen Uhlenbeck for her work in geometric analysis [20]. We thus propose that Sophie Germain's introduction of *la courbure moyenne* [9] defines her as a mathematician deserving of the highest attention for her mathematical vision and of the most profound recognition for her intellectual standing (for comparison, *see* also [4]).

Sophie Germain was a trailblazer, both as a female mathematician and as a differential geometer, introducing an important invariant that is used and generalized in today's geometrical theories. Her influence eased the way for other mathematical giants, such as Emmy Noether, or later, Michèle Audin, Dusa McDuff, and Chuu-Lian Terng, as well as the 2014 Fields Prize winner Maryam Mirzakhani. Following in her historical footsteps, we can see how these wonders of mathematics became inspiration for future generations of female mathematicians, may they be differential geometers, algebraists, or topologists.

To better assess the complexity of Sophie Germain's body of work in the context of her contributions to mathematical history as a female mathematician, we should compare it with those of other important cultural giants who played a singular part in their respective historical period. One such comparison could be made with Christine de Pizan (1364–1430), one of the first professional writers in medieval Europe, a biographer of King Charles V, and a first-hand witness of a historical period when her contemporary society descended into chaos and war. Barbero describes Christine de Pizan [1] as "the first feminist" and "*une femme engagée*", i.e., an independent intellectual who acts according to her principles and convictions while responding to the challenges of her time. If Christine de Pizan definitely is such an intellectual, exactly in the terms described by Barbero, then we should discuss Sophie Germain's imprint on the history of her times along the same lines, taking into account all aspects of her historical environment, from social prejudice to her life in the time of war and social tensions that would lead to permanent changes to French society. As in the case of Christine de Pizan, whose principles determined an attitude that today would be described as political, Sophie Germain did not hesitate to act in support of her values. For example, she did everything in her power, using all of her political influence, to protect Gauss when the French imperial army invaded his hometown (*see* her correspondence with General Pernety, who, in 1806, directed the siege of Breslau [10], pp. 316–317).

Jane Austen (1775–1817) was another contemporary of Sophie Germain. Similar to Germain, Jane Austen published anonymously, her name not appearing on her works until after her death. Her writing style presaged the literary realism movement, and the themes and political observations in her writing were so nuanced and important as to have legitimate claims by both conservatives and liberals. After comparing her appeal to that of Shakespeare and Dickens, Austen scholar John Mullan ([16], p. 2) writes that "… she did things with fiction that had never been done before. She did things with characterization, with dialogue, with English sentences, that had never been done before." It is not surprising, then, that there are hundreds of works of literary criticism devoted to Jane Austen's writings. We can only hope that Sophie's work will receive the same type of interest as Austen's literary and intellectual contributions.

While her mathematics was astounding, the scope of Sophie's intellectual brilliance is much wider. To better support our interest in all facets of her personality, we cite her volume of *Philosophical Works*, published in 1879 by Paul Ritti [10]. In particular, within this volume we refer the interested reader to a longer essay titled "General considerations on the state of sciences and of the letters in different times of their cultures," a series of short essays titled *Pensées*, as well as previously unreleased letters. Important information on her private correspondence was investigated and published only recently by A. Del Centina [6–8], doing justice to this interesting intellectual giant. All of these elements and texts should be taken into consideration when one discusses Sophie Germain's intellectual span and vision, and we feel that her intellectual life should be as important to the mathematical community as Jane Austen's intellectual vision is to the literary community. Despite the above-mentioned discussions of her philosophical works, we feel there is more to be done. Here, we briefly describe the main text in the volume titled by the editor *Philosophical Works*, i.e., the long essay "General considerations on the state of sciences and of the letters in different times of their cultures."

In the first chapter, Sophie Germain argues that, in various cultures, the development of sciences and the evolution of letters (including poetry and fiction) are governed by a common spirit, while in the second chapter, she starts by remarking that literature appeared in all world cultures before science. Her inquiry is not mathematical, and it definitely pertains to the philosophy of culture, as Sophie Germain is much interested in the origins of scientific inquiry, and this is best described in the following paragraph:

> *Les sciences n'existaient pas encore; mais le besoin d'expliquer s'était fait sentir. La première des littératures fut poétique. Ce qui tenait lieu des sciences physiques n'était pas moins poétique que la littérature elle-même ou plutôt ces deux branches du savoir, tellement séparées aujourd'hui qu'il faut de l'attention pour remarquer ce qu'elles ont du commun, étaient dans ces premiers temps entièrement confondues.*

(The sciences did not exist yet, but the need to explain was beginning to be felt. The first of the literatures was poetic. What took place in physical science was no less poetic than in literature itself, rather the two branches of knowledge, so far separate today that much attention is needed to identify what they have in common, were originally entirely entangled.) ([10] p. 113)

This transdisciplinary remark reveals not just the reflection of a research mathematician at work in the first decades of the nineteenth century, but a thorough historical vision. While it may be that her particular considerations are a product of the spirit of her time, it is important to point out that, by transcending the limitations of a single area of knowledge, most of Sophie Germain's essays exceed the vision and depth of most working mathematicians' reflections.

By the very fact that this mathematician, with important and numerous contributions to number theory, elasticity theory, and differential geometry, ventures into the territory of the philosophy of culture, we recognize that Sophie Germain is an authentic intellectual of her time, with a manifold interest in a variety of challenging ideas, who follows closely not only the current events and affairs of her era (e.g., the developments of the Napoleonic wars), but also their historical causalities. Although prejudices against females in academia prevented her from participating formally, she was very familiar with contemporary schools of thought and had a sophisticated perspective on the role of science in society. We also see that Sophie Germain had well-shaped opinions on a variety of scientists and their very specific work, as she also describes her preferences among them; she refers in her essay to a series of authors, some classics, such as Descartes and Newton, and some of her contemporaries, such as Immanuel Kant. She notes that

Newton parut, armé d'un nouveau genre de calcul: et l'unité, l'ordre, les proportions de l'univers que le sentiment du vrai avait fait chercher si longtemps devinrent des vérités mathématiques. Son génie avait reconnu la cause des mouvements célestes: une analyse pleine de finesse lui servit à les mesurer.

(Newton appeared, armed with a new kind of calculus: and the unity, order, and proportions of the universe, whose true reality had long been searched for, became mathematical truths. His genius recognized the cause of celestial movements: an analysis full of refinement served him to measure them.) ([10], p. 146.)

The unity of concepts was a modernist thought, which became highly valued a century later, and Sophie Germain points it out several times throughout her historical reflections; she seems to find, in Isaac Newton, a moral model and a more general example to follow. She notes:

En parlant de Newton qui fut solitaire et modeste, qui ne chercha point à paraître, qui fit des grandes choses avec simplicité, il faut être simple comme lui, comme la nature qu'il a suivie. Cette simplicité qui le charactérise est la grandeur que son écrivain doit emprunter de lui.

(Speaking of Newton, who was a loner and modest being, who did not seek to show off, who did great things with simplicity, one must be simple like him, like the nature he has followed. This simplicity which characterizes him is the greatness that any writer must borrow from him.) ([10], p. 258.)

Sophie Germain also reflects on what real life actually reserved for mathematicians during and before her time, and these reflections are as relevant today; one can feel in her ethical quest a reflection on her own destiny as a mathematician. She writes about others, but in many ways, she writes about herself when she says:

Tycho [Brahe] avait été destiné à la jurisprudence, comme Copernic le fut à la médecine.

(Tycho [Brahe] was destined to the legal profession, as Copernicus was to medicine.) ([10], p. 243.)

At some point, she seems to criticize Tycho Brahe for his lack of philosophical reflection ([10] pp. 247, 255), but she finds his attitude understandable, as he was a man much influenced by his century, where interests in alchemy merged with astronomical observations. By comparison, Sophie Germain has a much more positive take on Dominique Cassini's works and heritage ([10] pp. 256–257), whose works she finds "*précieux.*"

Sophie Germain is ultimately interested in what she sees as the fundamental duty of being a mathematician. Reminiscent of Hardy's *Apology* [11], she writes this reflection on the proper definition of a geometer:

Un géomètre est un homme qui entreprend de trouver la vérité, et cette recherche est toujours pénible dans les sciences comme dans la morale. Profondeur de vue, justesse de jugement, imagination vive, voilà les qualités du géomètre. Profondeur de vue pour apercevoir toutes les conséquences d'un principe, cette immense postérité d'un même père. Justesse de jugement, pour distinguer entre elles les traits de famille, et pour remonter de ces conséquences isolées au principe dont elles dépendent. Mais ce qui donne cette profondeur, ce qui exerce ce jugement, c'est l'imagination, non celle qui se joue à la surface des choses, qui les anime de ses couleurs, qui y répand l'éclat, la vie et le mouvement, mais une imagination qui agit au dedans des corps comme celle-ci au dehors. Elle se peint leur constitution intime, elle la change et la dépouille à volonté; elle fait, pour ainsi dire, l'anatomie des choses et ne leur laisse que les organes des effets qu'elle veut expliquer. L'une accumule pour embellir, l'autre divise pour connaître. L'imagination qui pénètre ainsi la nature, vaut bien celle qui tente de la parer. Moins brillante que l'enchanteresse qui nous amuse, elle a autant de puissance et plus de fidélité. Quand l'imagination a tout montré, les dif-ficultés et les moyens, le géomètre peut aller en avant; et s'il est parti d'un principe incontestable, qui rende sa solution certaine, on lui reconnaît un esprit sage. Ce principe le plus simple offre-t-il la voie la plus courte, il a l'élégance de son art. Et enfin il a du génie, s'il atteint une vérité grande, utile et longtemps séparée des vérités connues.

(A geometer is a man who undertakes to find the truth, and this research is always as painful in science as in morality. Depth of sight, correctness of judgment, lively imagination, these are the qualities of the geometer. Depth of sight to see all the consequences of a principle, this immense posterity from the same father. Judgment correctness, to distinguish the family traits between them, and to go back from these isolated consequences to the principle from which they spring. But what gives this depth, which exercises this judgment, is the imagination, not what is played on the surface of things, which animates them with its colors, which diffuses brightness, life, and movement, but an imagination that works just as well inside bodies as it does outside. It paints their intimate constitution; it changes it and strips it at will; it describes, so to speak, the anatomy of things, and leaves them only the organs of the effects which it wishes to explain. One accumulates for embellishment, the other divides for knowledge. The imagination that deeply penetrates nature is worth at least as much as the one that obscures it. Less brilliant than the enchantress who amuses us, she has as much power and more fidelity. When the imagination has shown everything, the difficulties and the means, the geometer can go forward; and if he has started from an incontestable principle, which renders certainty to his solution, he is recognized as having a wise mind. This simplest principle offers the shortest route; it has the elegance of its art. And finally he has genius, if he proves a great truth, useful and far removed from known truths.) ([10] pp. 266–267.)

Furthermore, any geometer would quantify the following fragment as one of the most interesting in her works, as it is premonitory and substantive in every sense:

La géométrie est la science de l'étendue et du mouvement ou seulement de l'étendue: car tout ce qui existe dans cet univers, ou à la fois ou successivement, a l'étendue pour caractère de son existence. L'espace qui embrasse tous les points, tous les lieux, toutes les bornes du physique; le mouvement qui parcourt cet espace, qui s'y applique, s'y mesure et semble s'y assimiler; le temps marqué par la succession des choses, subsistant depuis leur commencement jusqu'à leur fin; le temps qui embrasse l'univers dans ses changements, comme l'espace l'enferme dans sa permanence, tout n'est qu'étendue. Étendue physique qui est devant nous, que l'œil peut distinguer et parcourir, étendue intellectuelle que l'homme peut rendre présente à son esprit et qui n'est aperçue et mesurée que par la pensée. Voilà l'empire de la géométrie. C'est alors qu'elle est grande, qu'elle est vaste comme l'univers! Ouvrage miraculeuse de la raison humaine, les hommes y ont concentrée toutes les idées d'ordre et de rectitude, qu'ils ont reçues du ciel.

(Geometry is the science of magnitude and of movement, or only of magnitude, since all there is in this universe, either simultaneously or successively, has magnitude to characterize its existence. Space, which embraces all points, all places, all boundaries of the physical world; movement that passes through this space, which applies here, which is measured here, and is assimilated here; time marked by the succession of events, existing from their beginning up to their end; the time which embraces the universe in its changes, as well as the space confined in its eternity, all is nothing but magnitude. It is the physical magnitude that lies ahead, that our eyes can distinguish and cover, intellectual magnitude that man can spark in his spirit and which cannot be perceived and measured by anything else but by thought. That is geometry's empire. That is how large it stands, as wide as the whole universe! A wonderful miracle of human reason, people have focused inside all the ideas on order and on straightness they have received from the heavens.) ([10] pp. 262–263.)

Sophie Germain's interests pursued the fundamental principle to the ultimate realm, where she inventively resorts to effective metaphors to make her point:

La nature n'est que mélange et tempéraments, deux principes destructeurs l'un par l'autre enchaînés sont unis pour des effets durables. L'alliance de ces principes maintient la société des corps célestes! Rien n'est plus admirable que ce mécanisme, c'est par cette combinaison de forces que tout se meut, tout change et cependant tout se conserve!

(Nature is nothing else but mixture and disposition. Two principles destroying one another are interconnected to yield long-lasting consequences. The alliance of these principles keeps the combination of celestial bodies! Nothing is more admirable than this mechanism, due to this combination of forces that everything moves, everything changes, and at the same time everything is conserved.) ([10] pp. 258–259)

We would be remiss to forget Sophie Germain's note on human nature at a time when the Napoleonic wars left Europe devastated:

Nos moyens pour surpasser la science primitive ont donc été le télescope qui étend le domaine des sens, la géométrie qui permet de tout approfondir et le génie qui ose tout comparer et qui s'élève à la science des causes. Cette science est notre véritable supériorité. Tous les phénomènes sont enchaînés. Le système de nos connaissances est ordonné comme la nature; un seul principe nous sert à tout expliquer, comme un seul effort lui suffit pour faire tout agir.

(Our means to exceed primitive science have been the telescope, which extends the domain of the senses, geometry, which allows us to deepen everything, and the genius which dares to compare everything and which elevates to the science of causality. This science is our true superiority. All the phenomena are entangled. The system of our knowledge is as ordered as nature; one single principle serves us to explain everything, as a single effort is enough [to this principle] to make everything happen.) ([10] p. 281)

This paragraph is strongly reminiscent of Leo Tolstoy's concluding remarks from *War and Peace*, where the novelist is looking for the ultimate principles that govern major events such as the Napoleonic Wars. Tolstoy's masterpiece was published in its entirety in 1869, and this is how history felt in the nineteenth century. Consequently, we contend that Sophie Germain should be viewed not only as a research mathematician, but as a deep thinker, an intellectual facing and reflecting upon her time and on the forces of the natural world.

In the end, we are convinced that Sophie Germain feels most at home when she comments on mathematics, and we embrace her clear vision on relationships and entanglements between various chapters and concepts of mathematics. In this vein, Sophie Germain anticipates:

La méthode complète du calcul intégral serait une révolution dans la géométrie semblable à celle de l'application de l'algèbre et à celle de l'invention du calcul différentiel.

(The complete method of integral calculus would be a revolution in geometry similar to that of the applications of algebra and of differential calculus.) ([10] p. 281)

If we take into account her overall writings, her essays, and her private correspondence, Sophie Germain reveals herself as a fascinating scientist with an interesting humanistic personality, possessing eclectic interests, a very complex vision of mathematics and of the role of science in the world, as well as a personal vision of culture and philosophy, revealed in her vast array of reflections, composed in a unique and exquisite style. We can only speculate and wonder at what accomplishments such an active and brilliant mind would have achieved if Sophie Germain had been allowed to pursue her interests to their highest academic potential. In her destiny, there exists a historical lesson for us all. The history of mathematics simply does not have any other case of a researcher with such subtle and fundamental contributions, who faced a similar comprehensive system of prejudices and barriers, and who left such a transdisciplinary heritage. The historians of science, the translators, and the mathematicians who investigate her work perform a great service to the mathematical community.

We would like to end this well-deserved panegyric with Sophie Germain's own words. She writes the following in a poetical note that can only be described as a mark of her personal style for this entire genuine diary of ideas:

L'infini est le gouffre où se perdent nos pensées; il n'est pas naturel de se jeter dans des précipices. Si l'homme est descendu dans cet abîme sans fond, il y fut entraîné par une pente.

(The infinite is the chasm in which our thoughts are lost; it's not natural to throw oneself in its precipices. If the man descends in this endless abyss, he would be dragged into a fall.) ([10] p. 235.)

Acknowledgments

The last author would like to thank Professor Karen Uhlenbeck for her mathematical and life lessons, as they have induced growth and comfort in the author's life.

References

[1] A. Barbero. *Donne, Madonne, Mercanti e Cavalieri. Sei Storie Medievali*, Ed. Laterza, 2013.

[2] L. L. Bucciarelli and N. Dworsky, Sophie Germain. An essay in the history of the theory of elasticity, *Studies in the History of Modern Science*, 6. D. Reidel Publishing Co., Dordrecht-Boston, 1980.

[3] B.-Y. Chen. *Pseudo-Riemannian Geometry, δ-invariants and Applications*, World Scientific, 2011.

[4] A. Cogliati. Sulla ricezione del Theorema Egregium, 1828–1868, *Boll. Stor. Sci. Mat.* 38 (2018), no. 1, 31–60.

[5] A. Dahan-Dalmédico. Mécanique et théorie des surfaces: les travaux de Sophie Germain, *Historia Math.* 14 (1987), no. 4, 347–365.

[6] A. Del Centina. Letters of Sophie Germain preserved in Florence, *Historia Math.* 32 (2005), 60–75.

[7] A. Del Centina. Unpublished manuscripts of Sophie Germain and a revaluation of her work on Fermat's last theorem, *Arch. Hist. Exact Sci.* 62 (2008), no. 4, 349–392.

[8] A. Del Centina. The correspondence between Sophie Germain and Carl Friedrich Gauss, *Arch. Hist. Exact Sci.* 66 (2012), no. 6, 585–700.

[9] S. Germain. Mémoire sur la courbure des surfaces, *Journal für die reine und angewandte Mathematik*, 1831, pp. 1–29.

[10] S. Germain. Oeuvres philosophiques de Sophie Germain, suivies de pensées et de lettres inédites et précédées d'une notice sur sa vie et ses oeuvres par H. Stupuy, Paul Ritti, Paris, 1879.

[11] G. H. Hardy. *A Mathematician's Apology*, Cambridge: University Press, 1940; 2004.

[12] R. Laubenbacher and D. Pengelley. "Voici ce que j'ai trouvé:" Sophie Germain's grand plan to prove Fermat's last theorem, *Historia Math.* 37 (2010), no. 4, 641–692.

[13] G. Leibrock. Meine Freundin Sophie: Carl Friedrich Gauss' Brieffreundschaft mit Sophie Germain, *Gauss-Ges. Göttingen Mitt.* 38 (2001), 17–28.

[14] F. Marques and A. Neves. Min-Max theory and the Willmore conjecture, *Annals Math.* 179 (2014), 683–782.

[15] G. Micheli. The philosophical works of Sophie Germain (Italian), Science and philosophy, 712–729, Garzanti, Milan, 1985.

[16] J. Mullan. *What Matter in Jane Austen? Twenty Crucial Puzzles Solved*, Bloomsbury Press, 2012.

[17] J. H. Sampson. Sophie Germain and the theory of numbers, *Arch. Hist. Exact Sci.* 41 (1990), no. 2, 157–161.

[18] C. Truesdell. Sophie Germain: Fame earned by stubborn error, *Boll. Storia Sci. Mat.* 11 (1991), no. 2, 3–24.

[19] C. Truesdell. Jean-Baptiste-Marie Charles Meusnier de la Place (1754–1793): An historical note, *Meccanica*, 31 (1996), 607–610.

[20] K. Uhlenbeck. *Closed minimal surfaces in hyperbolic 3-manifolds, Seminar on minimal submanifolds*, Princeton University Press, 1983, pp. 147–168.

Who Owns the Theorem?

MELVYN B. NATHANSON

Simple questions: If you prove the theorem, do you own it? Can you forbid others to use or even cite it? Can you choose not to publish the theorem? Can you be forbidden to publish it?

What is a theorem? A mathematical statement may be true. It is true whether or not there is a proof. Without a proof, we do not know if it is true. A theorem is a true mathematical statement that has a proof.

Suppose there is a true mathematical statement, and you prove it. Now it is a theorem. It is "your theorem." In what sense might you own it? Can or should a theorem be considered the private property of its discoverer, who may or may not choose to publish? If you own the theorem, can you license it or rent it? Can you insist that anyone who wants to use or apply the theorem must pay you to do so?

If you publish the theorem in a refereed journal, or post it on arXiv, or explain it in a seminar, or submit it to a journal, then everyone knows you proved it. When does "your theorem" become part of the public library of proven mathematical truths that other researchers can freely use to prove new theorems?

If you need a result to prove a theorem, and know that the result is true but the discoverer has not announced or released it publicly, is it ethical (of course, properly citing the discoverer) to use that "unpublished" result in the proof? Is it ethical for you *not* to prove the theorem because it requires a result that is true but is being withheld by its "owner"?

Suppose you find out that someone has proved a theorem, but has not revealed it to the world. Maybe you have even seen the proof, and checked it, so you are sure that it is correct. Even though it has not been published, you know that it is a mathematical truth. Can you use it in a paper, even though the discoverer might not want the result to be known? Does the prover of the theorem own it enough to prevent other mathematicians from using it?

The notion of "owning a mathematical truth" is, in part, connected with careerism in academic life. What might be called "vulgar careerism" is endemic and not necessarily vulgar. Many mathematicians hide what they are working on so others will not "scoop" them, will not use "their" ideas to prove a theorem before they do. Perhaps it is not sufficient to give proper attribution. Maybe the author is an untenured assistant professor who wants to deduce more results from the theorem, publish more papers, and get promoted. Maybe the author thinks it will lead to a proof of the Riemann hypothesis and earn the million dollar prize from the Clay Mathematics Institute. Some mathematicians admit that they discuss their ideas about how to solve the Riemann hypothesis only after they are convinced that the ideas will not work.

It used to be that, every year, permanent professors in mathematics at the Institute for Advanced Study in Princeton would appoint a visiting member to be their "assistant." Long ago at the Institute, there was a permanent member who required his assistants to promise not to reveal to anyone what he was working on.[1] When I learned this, I was shocked. It was antithetical to everything I believed about science. I was also naive. I had not understood that for many people mathematics is a competition.[2]

In 1977, the National Security Agency (NSA) decided that publication of cryptographic research would endanger national security, and wanted to require that professors who wrote papers in cryptography would have to send them for pre-publication review by the NSA and not submit them to journals without NSA approval. At first, NSA hoped for voluntary compliance, but also considered making this a legal requirement. This did not happen, and after considerable contentiousness and debate in the mathematical community, prepublication review by the NSA faded away and seems not to be an issue today [1, 2].

Secrecy in mathematics is less important than in other sciences. Mathematical results rarely have commercial value. Like many mathematicians, I don't care if my theorems are "useful." I only hope that I have not made mistakes, that the proofs are correct, that the "theorems" are theorems and are interesting. I upload preprints to arXiv as soon as they are written, before I send them to a journal. I am happy if someone uses my results. But this does not answer the central question.

Mathematician A proves a theorem, and mathematician B learns about it. Maybe B reviewed A's NSF proposal. Maybe A submitted the manuscript to a journal and B refereed it. Can B use the theorem (as always, with proper attribution) in a paper before A has published it? For me, the answer is clear. Here is an analogy. Legally, you cannot sequence a plant or animal DNA strand and patent the sequence because you did not create the sequence. God created it, and you only discovered it. Similarly, mathematical truths exist, and mathematicians only discover them. If you discover a theorem, you have the power, the privilege, and, perhaps, the right not to reveal it to anyone, but if, somehow, someone learns of your result, knows that a certain mathematical statement is true, then that person has the right to tell the world and to apply it to obtain new results, with or without your consent.

Can you own a scientific truth? Can you hide a scientific truth? These are ethical questions, and, in the Covid era, not only in mathematics.

Notes

1. I was once André Weil's assistant at the Institute. He did not impose a secrecy oath.

2. I still do not understand why, for some mathematicians, getting medals in high school and college competitions is a core part of their self-esteem.

References

[1] S. Landau, *Primes, codes, and the National Security Agency*, Notices Amer. Math. Soc. 30 (1983), 7–10.

[2] W. Diffie and S. Landau, *Privacy on the Line, Updated and Expanded Edition: The Politics of Wiretapping and Encryption*, MIT Press, Cambridge, MA, 2007.

A Close Call:
How a Near Failure
Propelled Me to Succeed

TERENCE TAO

For as long as I can remember, I was always fascinated by numbers and the formal symbolic operations of mathematics, even before I knew the uses of mathematics in the real world. One of my earliest childhood memories was demanding that my grandmother, who was washing the windows, put detergent on the windows in the shape of numbers. When I was particularly rowdy as a child, my parents would sometimes give me a math workbook to work on instead, which I was more than happy to do. To me, mathematics was an activity to do for fun, and I would play with it endlessly.

Perhaps because of this, I found my mathematics classes at school to be easy—perhaps too easy—even after skipping a number of grades. If a lecture was on a topic I found interesting, I would use the class time to experiment with the material, perhaps finding alternate derivations of some step the teacher did on the board, or to plug in some numbers to try out special cases and look for patterns. If instead I found the topic to be dull, I would doodle like any other bored student. In either case, I did not take particularly detailed notes, nor did I ever develop any systematic study habits. I would be able to improvise my way through my homework and exams, for instance, by cramming through the textbook a few days before a final exam and perhaps playing a bit more with the parts of the class material that I really liked. It tended to work fairly well all the way up to my undergraduate classes. The courses that I enjoyed, I aced; classes that I found boring, I only barely passed, or (in two cases) failed altogether. (One class was a FORTRAN programming class in which I had refused to learn FORTRAN on the grounds that

I already knew how to program in BASIC; the other was a quantum mechanics class in which we were warned well ahead of time that the final exam would require us to write a short essay on the history of the subject, which I totally ignored until the day of the exam, during which I still recall having to be escorted from the examination room in tears.) Despite this, I ended up graduating from my university with honors at the top of my class—but it was a small university with a tiny honors program, and in fact there were only two other honors students in mathematics in my year!

When I entered graduate study at Princeton, I brought my study habits (or lack thereof) with me. At the time in Princeton, the graduate classes did not have any homework or tests; the only major examination one had to pass (apart from some fairly easy language requirements) were the dreaded "generals"—the oral qualifying exams, often lasting more than two hours, that one would take in front of three faculty members, usually in one's second year. The questions would be drawn from five topics: real analysis, complex analysis, algebra, and two topics of the student's choice. For most of the other graduate students in my year, preparing for the generals was a top priority; they would read textbooks from cover to cover, organize study groups, and give each other mock exams. It had become a tradition for every graduate student taking the generals to write up the questions they received and the answers they gave for future students to practice. There were even skits performed (with much gallows humor) on hypothetical general exams with a "death committee" of three faculty that were particularly notorious for being harsh on the examinee.

I managed to brush off almost all of this. I went to the classes that I enjoyed, dropped out of the ones I did not, and did some desultory reading of textbooks but spent an embarrassingly large fraction of my early graduate years messing around online (having discovered the World Wide Web in my first year) or playing computer games until late at night at the graduate dormitory computer room. For my general topics, I chose harmonic analysis—which I had studied for my master's degree back in Australia—and analytic number theory. Feeling that analysis was my strong suit, I only spent a few days reviewing real, complex, and harmonic analysis; the bulk of my study, such as it was, was devoted instead to algebra and analytic number theory. All in all, I probably only did about two weeks' worth of preparation for the generals, while my fellow classmates had devoted months. Nevertheless, I felt quite confident going into the exam.

The exam started off reasonably well, as they asked me to present the harmonic analysis that I had prepared, which was mostly material based on my master's thesis and specifically on a theorem in harmonic analysis known as the T(b) theorem. However, as they moved away from that topic, the shallowness of my preparation in the subject showed quite badly. I would be able to vaguely recall a basic result in the field, but not state it accurately, give a correct proof, or describe what it was used for or connected to. I have a distinct memory of the examiners asking easier and easier questions, to get me to a point where I would actually be able to give a satisfactory answer; they spent several minutes, for instance, painfully walking me through a derivation of the fundamental solution for the Laplacian. I had enjoyed playing with harmonic analysis for its own sake and had never paid much attention as to how it was used in other fields, such as partial differential equations (PDEs) or complex analysis. Presented, for instance, with the Fourier multiplier for the propagator of the wave equation, I did not recognize it at all and was unable to say anything interesting about it.

At this point, I was saved by a stroke of pure luck, as the questioning then turned to my other topic of analytic number theory. Only one of the examiners had an extensive background in number theory, but he had mistakenly thought I had selected algebraic number theory as my topic, and so all the questions he had prepared were not appropriate. As such, I only got very standard questions in analytic number theory (e.g., prove the prime number theorem, Dirichlet's theorem, etc.), and these were topics that I actually did prepare for, so I was able to answer these questions quite easily. The rest of the exam then went fairly quickly, as none of the examiners had prepared any truly challenging algebra questions.

After many nerve-wracking minutes of closed-door deliberation, the examiners did decide to (barely) pass me; however, my advisor gently explained his disappointment at my performance and told me I needed to do better in the future. I was still largely in a state of shock—this was the first time I had performed poorly on an exam that I was genuinely interested in performing well in. But it served as an important wake-up call and a turning point in my career. I began to take my classes and studying more seriously. I listened more to my fellow students and other faculty, and I cut back on my gaming. I worked particularly hard on all of the problems that my advisor gave me, in the hopes of finally impressing him. I certainly did not always succeed at this—for instance, the first problem my advisor gave me, I was only able to solve five years after my Ph.D.—but I poured substantial effort into the last two years of my graduate study, wrote up a decent thesis and a number of publications, and began the rest of my career as a professional mathematician. In retrospect, nearly failing the generals was probably the best thing that could have happened to me at the time.

My write-up of my general exam experience is still available online. I have been told that it has been a significant source of comfort to the more recent graduate students at Princeton.

Contributors

Mike Askew is Distinguished Professor of Mathematics Education at the University of Witwatersrand, Johannesburg, before which he held Chair Professorships at Monash University, Melbourne, and King's College, University of London, where he obtained his M.A. (1988) and Ph.D. (2000), with a B.Sc. (1975) in Pure Mathematics from the University of Sheffield. He researches how to improve the teaching of school mathematics, and his more than 200 publications largely focus on the central importance of reasoning and problem solving in teaching and learning mathematics. From 2018 to 2019, he was President of the United Kingdom's Mathematical Association.

Michael J. Barany (http://orcid.org/0000-0002-4067-5112) researches and teaches the history, sociology, and culture of science and mathematics at the University of Edinburgh. He co-edited, with Kirsti Niskanen, the edited volume *Gender, Embodiment, and the History of the Scholarly Persona* (Palgrave 2021) and has two monographs in progress: an academic book on the globalization of modern mathematics and a trade book on how to understand mathematics in social context. This is his fifth appearance in the *Best Writing on Mathematics* series. You can read more of his work at http://mbarany.com and follow him on twitter at @mbarany.

Viktor Blåsjö is an assistant professor at the Mathematical Institute of Utrecht University. As a historian with a mathematical home base and a predilection for geometry, he addresses issues in philosophy, the history of science, and historiography from a perspective informed by mathematical content. You can follow him on Twitter at @viktorblasjo and listen to his Opinionated History of Mathematics podcast.

John H. Conway passed away of coronavirus complications in 2020. He was the John von Neumann Distinguished Professor of Mathematics at Princeton University, and he embodied the playful spirit in mathematics. He worked in many areas of mathematics, notably group theory, knot theory, game theory, and geometry. He is best known for his work on the Game of Life and the Free Will Theorem, but the contribution of which he was most proud was the discovery of the surreal numbers. He was always happy to engage in any kind of mathematical game, including Go, Phutball (his own invention), or even Dots and Boxes.

Michael C. Duddy is Assistant Professor of Architecture at New York City College of Technology, City University of New York. He holds a B.A. from New York University and a Master of Architecture from the Yale School of Architecture. His writings include "The Ends of Reason: Towards an Understanding of the Architectonic," *The Journal of Aesthetics and Phenomenology* (Routledge 2018) and "Roaming Point Perspective: A Dynamic Interpretation of the Visual Refinements of the Greek Doric Temples," *Nexus Network Journal* (Springer 2008). His research interests explore the intersection of Husserlian phenomenology and Kantian epistemology, and its presentation in architectural form.

The late **Harold M. Edwards** (1936–2020) was a professor emeritus of mathematics at New York University. He received the Whiteman and Steele prizes of the American Mathematical Society and was the author of several books, including *Advanced Calculus, Riemann's Zeta Function, Fermat's Last Theorem, Galois Theory, Divisor Theory, Linear Algebra, Essays in Constructive Mathematics,* and *Higher Arithmetic.*

Stephan Ramon Garcia is W.M. Keck Distinguished Service Professor at Pomona College. He is the author of four books and more than 100 research articles in operator theory, complex analysis, matrix analysis, number theory, discrete geometry, combinatorics, and other fields. He has served on the editorial boards of the *Proceedings of the American Mathematical Society*, *Notices of the American Mathematical Society, Involve,* and *The American Mathematical Monthly.* He has received four National Science Foundation research grants as principal investigator, and six teaching awards from three different institutions. In 2019, he earned the inaugural Dolciani Prize for Excellence in Research and was elected a Fellow of the American Mathematical Society

Adam Glesser earned his Ph.D. in Mathematics at the University of California, Santa Cruz, working with Robert Boltje. He is now a Professor at California State University, Fullerton. His research area is finite group representation theory. His work has appeared in journals such as *Transactions of the American Mathematical Soc*iety, *Proceedings of the American Mathematical Soc*iety, *Journal of the London Mathematical Society,* and he received the 2020 Pólya

Award from the Mathematical Association of America for an article in *The College Mathematics Journal*. Outside of mathematics, Adam is an avid board gamer, a father of three, and a minimalist hoarder.

David J. Hand (https://orcid.org/0000-0002-4649-5622) is Emeritus Professor of Mathematics and a Senior Research Investigator at Imperial College, London. He is a Fellow of the British Academy and a former President of the Royal Statistical Society. He has received the Guy Medal of the Royal Statistical Society, the Box Medal from the European Network for Business and Industrial Statistics, and the International Research Medal of the International Federation of Classification Societies. His 31 books include *Principles of Data Mining*, *Measurement Theory and Practice*, *The Improbability Principle*, *Statistics: A Very Short Introduction*, *The Wellbeing of Nations*, and *Dark Data*.

Kevin Hartnett is the senior math writer for *Quanta Magazine*. Previously he wrote "Brainiac," a weekly column for the *Boston Globe*'s Ideas section. A native of Maine, Kevin lives in South Carolina with his wife and three sons. Follow him on Twitter @kshartnett.

From 1974 to 2015, **Roger Howe** taught and did mathematical research at Yale University. His research focuses on symmetry, especially the notion of a "dual pair" of subgroups of a group, which links the areas of invariant theory, representation theory, automorphic forms, and physics. Since 1990, he has been increasingly occupied with issues of mathematics education. In 1999–2001, he served on the National Research Council's Mathematics Learning Study Committee that produced the report "Adding It Up." In 2015, he retired from Yale and joined the Department of Teaching, Learning & Culture at Texas A&M University, to lead a project to improve the mathematical preparation of elementary school teachers. He was a Guggenheim Fellow in 1982–1983. He is a Fellow of the American Mathematical Society and a member of the National Academy of Sciences and the American Academy of Arts and Sciences.

Andrew Lewis-Pye is a Professor in the Department of Mathematics at the London School of Economics (LSE). His research interests are in computability, algorithmic randomness, network science, and cryptocurrencies. Before his work at LSE, he was a Royal Society University Research Fellow at the University of Leeds and a Marie-Curie Fellow at the University of Siena.

Ben Logsdon (https://orcid.org/0000-0002-9791-7178) recently graduated from Williams College with a degree in mathematics and computer science. He is currently pursuing a Ph.D. in mathematics at Dartmouth College. His research interests include number theory and algebra, as well as applying ideas from mathematical logic and category theory to the design of programming languages and proof assistants.

Stephen K. Lucas (https://orcid.org/0000-0003-1827-7086) is a Professor in the Department of Mathematics and Statistics at James Madison University. His main research interests are computational, but they range across areas including differential equations, computational number theory, graph theory, the history of computation, and number representations. In 2002, he received the JH Michell Medal for outstanding new researchers in applied mathematics in Australia.

Sanjoy Mahajan (https://orcid.org/0000-0002-9565-9283) studied mathematics at Oxford University as a Marshall Scholar, received his Ph.D. in physics at Caltech, and is interested in the art of approximation and physics education. He has taught varying subsets of physics, mathematics, electrical engineering, and mechanical engineering at the African Institute for Mathematical Sciences, the University of Cambridge, and MIT, where he is in Open Learning and the Mathematics Department. He is the author of *Street-Fighting Mathematics* (MIT Press 2010), *The Art of Insight in Science and Engineering* (MIT Press 2014), and *A Student's Guide to Newton's Laws of Motion* (Cambridge University Press 2020).

Anya Michaelsen (https://orcid.org/0000-0002-5082-0184) earned her B.A. in Mathematics from Williams College in 2019 and is currently an NSF Graduate Research Fellow in Mathematics at UC Berkeley. After earning an Outstanding Presentation Award (MathFest, 2018) for presenting her undergraduate thesis "Uncountable n-Dimensional Excellent Regular Local Rings with Countable Spectra," she authored "Design and Construction of Mathematical Posters" for the 2020 *Notices of the American Mathematical Society* Early Career section. Throughout her time at Williams and Berkeley, she has been deeply engaged in department culture through student groups and department committees, with a focus on equity and inclusion, as well as support systems for all students.

Don Monroe received a Ph.D. in experimental condensed-matter physics from MIT and spent the next 18 years in research at Bell Labs. In 2003, he returned to school to get a Master's degree from NYU's science writing program. He has written on a range of topics from systems biology and cancer genetics to quantum computing and integrated-circuit technology, for publications including *Physics: Focus, Scientific American, Technology Review, New Scientist, Science*, publications of The New York Academy of Sciences, and *Communications of the Association of Computing Machinery*. An up-to-date list of his work is at http://www.DonMonroe.info.

Ralph Morrison (https://orcid.org/0000-0001-7134-1521) is an assistant professor of mathematics at Williams College who works in tropical geometry, with a special focus on tropical curves and chip-firing games on graphs. He received a B.A. in mathematics from Williams College in 2010 and a Ph.D. in mathematics from UC Berkeley in 2015, where his dissertation received the Kenneth Ribet & Lisa Goldberg Award in Algebra. He is actively involved in the mentorship of undergraduates, particularly through Williams College's SMALL REU program in mathematics and through the Mentor Project. He contributed the chapter "Tropical Geometry" to the book *A Project-Based Guide to Undergraduate Research in Mathematics*.

U. S. S. R. Moscow is the capital of Russia, a major urban center with a population of about 15 million residents and a city with almost 900 years of history. It is the northernmost and the coldest megacity on Earth. Moscow is the largest science center in Russia, the home of the Russian Academy of Sciences and of its numerous research institutes. It has roughly 200 institutions of higher education, including the flagship Lomonosov Moscow State University, founded in 1755.

Yelda Nasifoglu (https://orcid.org/0000-0003-3854-2035) is a historian of early modern mathematics and architecture and an Associate Member of the Faculty of History, University of Oxford. She received her doctorate in the history and theory of architecture from McGill University with her dissertation, "Robert Hooke's *Praxes*: Reading, Drawing, Building" (2018), and has been a researcher with the American Human Rights Council-funded project Reading Euclid's *Elements of Geometry* in Early Modern Britain and Ireland, based at Oxford. Most recently, she co-edited the volume of essays *Reading Mathematics in Early Modern Europe* (2020) and is developing a digital humanities project on the mathematical book trade in early modern Britain.

Melvyn B. Nathanson (http://orcid.org/0000-0002-6191-9374) is a professor of mathematics at the City University of New York (Lehman College and the CUNY Graduate Center) and a Fellow of the American Mathematical Society. He studied philosophy at the University of Pennsylvania and biophysics at Harvard University before completing his Ph.D. in mathematics at the University of Rochester in 1971. He worked with I. M. Gel'fand at Moscow

State University under an IREX fellowship in 1972–1973 and was Assistant to André Weil at the Institute for Advanced Study in 1974–1975. He has had visiting appointments at Harvard University, Princeton University, and Tel Aviv University. This is his fourth appearance in *The Best Writing in Mathematics* series.

M. S. Paterson (Ph.D., FRS) took degrees in mathematics at Cambridge University and rose to fame as the co-inventor with John Conway of Sprouts. His lifelong fascination with computer science began in the mid-1960s with the arrival in Cambridge of a massive new computer with up to 1 megabyte of memory. He evolved from president of the Trinity Mathematical Society to president of the European Association for Theoretical Computer Science and migrated from MIT to the University of Warwick, where he has been in the Computer Science Department for 48 years.

Steve Pomerantz has a B.A. from Queens College (1981) and a Ph.D. in Mathematics from UC Berkeley (1986) specializing in partial differential equations. He is a mathematical consultant and teaches math part-time in New York City. A long private career steeped in partial differential equations and probability has transformed to the study of his first passion: geometry. He is also an artist specializing in paintings inspired by the geometric art of world faiths. He is actively involved in the Monterey Bay Area Math Project, providing workshops in a variety of mathematical-art areas as a complement to the teaching of geometry. Notes from these workshops are available on his website circleofsteve.com and have been published in *Classical Geometry: An Artistic Approach*.

Evelyn Sander (https://orcid.org/0000-0003-4478-3919) is a Professor in the Department of Mathematical Sciences at George Mason University, working in the fields of dynamical systems and mathematical design and visualization. She developed the George Mason capstone course "Mathematics through 3D printing." She previously served as Section Editor of *SIAM Review*'s Research Spotlights section and the Editor-in-Chief of *DSWeb Magazine*, and now serves as the Editor-in-Chief of the *SIAM Journal on Applied Dynamical Systems*. She regularly mentors undergraduate research, and she has graduated five Ph.D. students.

Jacob Siehler is an associate professor of mathematics at Gustavus Adolphus College and a recipient of the George Pólya Award for his article "The Finite Lamplighter Groups: A Guided Tour" in *The College Mathematics Journal*. He has taught mathematics at every level from first grade to university, including one particularly gratifying summer in New York working with seventh-grade students in the Bridge to Enter Advanced Mathematics (BEAM) program.

Joshua Sokol is a freelance science writer based in Raleigh, North Carolina. He holds a B.A. in Astronomy and English Literature from Swarthmore College and an S.M. in Science Writing from MIT. His work, published in outlets like *Science*, *Quanta Magazine*, and *The New York Times*, has previously been recognized by the American Astronomical Society, the Council for the Advancement of Science Writing, the American Institute of Physics, the American Geophysical Union, and The Best American Science and Nature Writing series.

Bogdan D. Suceavă (https://orcid.org/0000-0003-3361-3201) earned his doctoral degree at Michigan State University in 2002, under Professor Bang-Yen Chen's coordination. His works appeared in the *Houston Journal of Mathematics*, the *Taiwanese Journal of Mathematics, American Mathematical Monthly, Mathematical Intelligencer, Differential Geometry and Its Applications, Czechoslovak Mathematical Journal, Results in Mathematics, Notices of the American Mathematical Society*, and *Historia Mathematica*, among other journals. He is Professor at California State University, Fullerton. Recipient of the 2020 Pólya Award of the Mathematical Association of America for a work co-authored with A. Glesser, M. Rathbun, and I. M. Serrano. This is the third time his work is featured in *The Best Writing on Mathematics*.

Laura Taalman is a Professor of Mathematics at James Madison University whose published research has included algebraic geometry, knot theory, and games. Also known as "mathgrrl," Dr. Taalman is a computational designer who leverages a diverse toolbox of 3D design software and technical materials to create elegant and aesthetic realizations of idealized mathematical objects. She is a Project NExT Fellow, a recipient of the Alder Award, the Trevor Evans Award, and the SCHEV Outstanding Faculty Award, and has been featured on Thingiverse, Adafruit, and National Public Radio's Science Friday.

Terence Tao (https://orcid.org/0000-0002-0140-7641) was born in Adelaide, Australia, in 1975. He has been a professor of mathematics at UCLA since 1999, having completed his Ph.D. under Elias Stein at Princeton in 1996. Tao's areas of research include harmonic analysis, partial differential equations, combinatorics, and number theory. He has received a number of awards, including the Salem Prize in 2000, the Fields Medal in 2006, the MacArthur Fellowship in 2007, and the Breakthrough Prize in Mathematics in 2015. Terence Tao also currently holds the James and Carol Collins Chair in Mathematics at UCLA.

Maria Trnkova (https://orcid.org/0000-0002-7994-5749) is a lecturer at the University of California in Davis. She obtained her Ph.D. from Palacky University, Czech Republic, in 2012. The main part of her thesis was based on a collaborative work with David Gabai "Exceptional hyperbolic three-manifolds" (*Commentarii Mathematici Helvetici*). Her area of research is low-dimensional geometry and topology, but she is also interested in applied mathematics. Her recent work with Joel Hass, "Approximating Iso-surfaces by guaranteed-quality triangular meshes," was published by the Eurographics Symposium on Geometry Processing 2020.

Mihaela B. Vâjiac (https://orcid.org/0000-0003-1526-5952) defended her doctoral thesis at Boston University under the coordination of Steve Rosenberg, then worked at the University of Texas at Austin as a postdoctoral associate under the supervision of Professor Karen Uhlenbeck. Presently she serves as Associate Professor of Mathematics and Director of the Center of Excellence in Complex and Hypercomplex Analysis at Chapman University. Her research interests include complex and hypercomplex analysis, operator theory, differential geometry, and integrable systems. Her papers have appeared in *Journal of Differential Geometry, Mathematische Nachrichten, Houston Journal of Mathematics, The Journal of Geometric Analysis*, and other journals.

Stan Wagon (http://stanwagon./com) (https://orcid.org/0000-0002-4524 -0767) obtained his Ph.D. in set theory at Dartmouth College in 1975 and is now retired from teaching, having taught at Smith and Macalester Colleges. He has written more than 100 papers and many books and has received many writing awards. The only bicycle he ever ridden is a bizarre full-sized square-wheeled bicycle; that attracted a lot of attention and earned him an entry in Ripley's *Believe It or Not*. His books include *The Banach-Tarski Paradox* and two bicycle-related books: *Which Way Did the Bicycle Go?* and *Bicycle or Unicycle?* He has been a runner and long-distance skier for many years and was a founding editor of *UltraRunning* magazine.

Michael Wallace (https://orcid.org/0000-0002-5763-5723) is an assistant professor of biostatistics at the University of Waterloo in Ontario, Canada. They completed their undergraduate degree in mathematics at the University of Cambridge, before an M.Sc. in statistics at University College London and a Ph.D.—also in statistics—at the London School of Hygiene and Tropical Medicine. Their research centers on precision medicine, where treatment decisions are tailored to patient-level characteristics. A key aspect to this work lies in communication of statistical ideas to a wide variety of audiences. Follow them on Twitter at @statacake.

Natalie Wolchover is a senior writer and editor at *Quanta Magazine*, where she primarily covers the physical sciences. She also has bylines in *Nature*, *NewYorker.com*, *Popular Science,* and other publications. Her writing has been featured in *The Best American Science and Nature Writing* and *The Best Writing on Mathematics* and has won awards including the 2016 Evert Clark/Seth Payne Award for young science journalists and the American Institute of Physics' 2017 Science Communication Award. Natalie has a bachelor's degree in physics from Tufts University, where she co-authored several papers in nonlinear optics. You can follow her on Twitter at @nattyover.

Notable Writings

Several interesting volumes besides this one can be put together by choosing from the enormous amount of literature on mathematics published last year. In this section you can find such suggestions.

As a space-saving rule, the two lists that follow are complementary; that is, I did not include in the first list the titles of articles published in the special journal issues mentioned on the second list. Because of severe time and other resource constraints related to the coronavirus health crisis, I did not cover as much literature as I would have liked. Even in normal times, some periodicals are available to me only in paper copies, in the libraries at Cornell University and Syracuse University; with both libraries closed while I worked on this section, my bibliographic research was negatively impacted.

In some electronic-only journals, the pagination of articles no longer starts with the page number that follows the last page of previous articles—at least in the versions I downloaded. In such cases, I gave the pagination starting with page 1 (one) and ending with the page number equal to the number of pages of the respective piece. Thus, occasionally, different entries appear to overlap in reference location.

Notable Journal Articles

Abrams, Ellen. "'An Inalienable Prerogative of a Liberated Spirit': Postulating American Mathematics." *British Journal for the History of Mathematics* 35.3(2020): 225–245.

Abrams, Ellen. "'Indebted to No One': Grounding and Gendering the Self-Made Mathematician." *Historical Studies in the Natural Sciences* 50.3(2020): 217–247.

Agirbas, Asli. "Algorithmic Decomposition of Geometric Islamic Patterns: A Case Study with Star Polygon Design in the Tombstones of Ahlat." *Nexus Network Journal* 22.1(2020): 113–137.

Alassi, Sepideh. "Jacob Bernoulli's Analyses of the *Funicularia* Problem." *British Journal for the History of Mathematics* 35.2(2020): 137–161.

Aleotti, Sara, Francesco Di Girolamo, Stefano Massaccesi, and Konstantinos Priftis. "Numbers around Descartes: A Preregistered Study on the Three-Dimensional SNARC [Spatial-Numerical Association of Response Codes] Effect." *Cognition* 195(2020): 1–11.

Andersen, Line Edslev. "Acceptable Gaps in Mathematical Proofs." *Synthese* 197.1(2020): 233–247.

Artzrouni, Marc. "Are Models Useful? Reflections on Simple Epidemic Projection Models and the Covid-19 Pandemic." *The Mathematical Intelligencer* 42.3(2020): 1–9.

Bach, Joscha. "Don't 'Flatten the Curve,' Squash It!" *Medium* Mar 13, 2020. https://medium.com/@joschabach/flattening-the-curve-is-a-deadly-delusion-eea324fe9727.

Bagley, Spencer. "The Flipped Classroom, Lethal Mutations, and the Didactical Contract: A Cautionary Tale." *PRIMUS* 30.3(2020): 243–260.

Bair, Jacques, Piotr Błaszczyk, Elías Fuentes Guillén, Peter Heinig, Vladimir Kanovei, and Mikhail G. Katz. "Continuity between Cauchy and Bolzano: Issues of Antecedents and Priority." *British Journal for the History of Mathematics* 35.3(2020): 207–224.

Bakker, Arthur, and David Wagner. "Pandemic: Lessons for Today and Tomorrow?" *Educational Studies in Mathematics* 104(2020): 1–4.

Bandt, Christoph, and Dmitry Mekhontsev. "Computer Geometry: Rep-Tiles with a Hole." *The Mathematical Intelligencer* 42.1(2020): 1–5.

Baron, Sam, Mark Colyvan, and David Ripley. "A Counterfactual Approach to Explanation in Mathematics." *Philosophia Mathematica* 28.1(2020): 1–34.

Basyal, Deepak. "A Mathematical Poetry Book from Nepal." *British Journal for the History of Mathematics* 35.3(2020): 189–206.

Belyaev, Alexander, and Pierre-Alain Fayolle. "Polygon Offsetting with Squares Erected on Its Sides." *The Mathematical Intelligencer* 42.4(2020): 38–41.

Berghofer, Philipp. "Intuitionism in the Philosophy of Mathematics: Introducing a Phenomenological Account." *Philosophia Mathematica* 28.2(2020): 204–235.

Biagioli, Francesca. "Ernst Cassirer's Transcendental Account of Mathematical Reasoning." *Studies in History and Philosophy of Science, Part A* 79(2020): 30–40.

Bø, Erlend E., Elin Halvorsen, and Thor O. Thoresen. "Investigating the 'Carnegie Effect'." *Significance* 17.3(2020): 6–7.

Bolondi, Giorgio, Federica Ferretti, and Andrea Maffia. "Monomials and Polynomials: The Long March towards a Definition." *Teaching Mathematics and Its Applications* 39.1(2020): 1–12.

Bortot, Alessio, and Agostino De Rosa. "Warped Curves in the Vestibule Arch of the Pio Clementino Museum, Rome." *Nexus Network Journal* 22.1(2020): 9–24.

Boulesteix, AnneLaure, Sabine Hoffmann, Alethea Charlton, and Heidi Seibold. "A Replication Crisis in Methodological Research?" *Significance* 17.5(2020): 18–21.

Boyce, Kenneth. "Mathematical Application and the No Confirmation Thesis." *Analysis* 80.1(2020): 11–20.

Bréard, Andrea, and Constance A. Cook. "Cracking Bones and Numbers: Solving the Enigma of Numerical Sequences on Ancient Chinese Artifacts." *Archive for History of Exact Sciences* 74.4(2020): 313–343.

Bruderer, Herbert. "The Antikythera Mechanism." *Communications of the ACM* 63.4(2020): 108–115.

Brueckler, Franka Miriam, and Vladimir Stilinović. "An Early Appearance of Nondecimal Notation in Secondary Education." *The Mathematical Intelligencer* 42.3(2020): 50–54.

Büscher, Christian. "Scaling Up Qualitative Mathematics Education Research through Artificial Intelligence Methods." *For the Learning of Mathematics* 40.2(2020): 2–7.

Cabeleira, João. "Deconstructing the Imaginary Space of a Quadratura." *Nexus Network Journal* 22.1(2020): 25–44.

Calvillo, Gonzalo Sotelo. "Tessellations in the Architecture of Pablo Palazuelo." *Nexus Network Journal* 22.2(2020): 349–368.

Campbell, J. M. "On the Visualization of Large-Order Graph Distance Matrices." *Journal of Mathematics and the Arts* 14.4(2020): 297–330.

Capaldi, Mindy. "What Definitions Are Your Students Learning?" *PRIMUS* 30.4(2020): 400–414.

Capanna, Alessandra. "The House in Four Dimensions Is a Theorem." *Nexus Network Journal* 22.1(2020): 45–59.

Carlevaris, Laura. "N-Dimensional Space and Perspective: The Mathematics behind the Interpretation of Ancient Perspective." *Nexus Network Journal* 22(2020): 601–614.

Carson, Cathryn. "Clouds of Data." *Historical Studies in the Natural Sciences* 50.1–2(2020): 81–89.

Cellucci, Carlo. "The Role of Notations in Mathematics." *Philosophia* 48(2020): 1397–1412.

Çetinkaya-Rundel, Mine, and Maria Tackett. "From Drab to Fab: Teaching Visualization via Incremental Improvements." *Chance* 33.2(2020): 31–41.

Chai, Christine P. "The Importance of Data Cleaning: Three Visualization Examples." *Chance* 33.1(2020): 4–9.

Chakravarthi, Ramakrishna, and Marco Bertamini. "Clustering Leads to Underestimation of Numerosity, but Crowding Is Not the Cause." *Cognition* 198(2020): 1–15.

Chu, Junyi, Pierina Cheung, Rose M. Schneider, Jessica Sullivan, David Barner. "Counting to Infinity: Does Learning the Syntax of the Count List Predict Knowledge That Numbers Are Infinite?" *Cognitive Science* 44.8(2020): 1–30.

Clivaz, Stéphane, and Takeshi Miyakawa. "The Effects of Culture on Mathematics Lessons: An International Comparative Study of a Collaboratively Designed Lesson." *Educational Studies in Mathematics* 105(2020): 53–70.

Cochran, James J. "Pandemics and Exponential Growth." *Significance* 17.3(2020): 18–21.

Craik, Alex D. D. "Henry Parr Hamilton (1794–1880) and Analytical Geometry at Cambridge." *British Journal for the History of Mathematics* 35.2(2020): 162–170.

Czocher, Jennifer A., and Keith Weber. "Proof as a Cluster Category." *Journal for Research in Mathematics Education* 51.1(2020): 50–74.

Danesi, Marcel. "Mathematical Cognition and the Arts." *Journal of Humanistic Mathematics* 10.1(2020): 317–336.

Davis, Tara C., and Anneliese H. Spaeth. "Using Reading Journals in Calculus and Beyond." *PRIMUS* 30.7(2020): 814–826.

Dawson, Robert, and Pietro Milici. "Rectification of Circular Arcs by Linkages." *The Mathematical Intelligencer* 42.1(2020): 18–23.

Del Centina, Andrea. "Pascal's *Mystic Hexagram*, and a Conjectural Restoration of His Lost Treatise on Conic Sections." *Archive for History of Exact Sciences* 74.5(2020): 469–521.

Del Centina, Andrea, and Alessandra Fiocca. "Borelli's Edition of Books V–VII of Apollonius's *Conics*, and Lemma 12 in Newton's *Principia*." *Archive for History of Exact Sciences* 74.3(2020): 255–279.

Diggle, Peter J., Tim Gowers, Frank Kelly, and Neil Lawrence. "Decision-Making with Uncertainty." *Significance* 17.6(2020): 12.

Divjak, Dagmar, and Petar Milin. "Exploring and Exploiting Uncertainty: Statistical Learning Ability Affects How We Learn to Process Language along Multiple Dimensions of Experience." *Cognitive Science* 44.5(2020): 1–32.

Dobie, Tracy E., and Miriam Gamoran Sherin. "What's in a Name? Language Use as a Mirror into Your Teaching Practice." *The Mathematics Teacher* 113.5(2020): 354–360.

Dotan, Dror, and Stanislas Dehaene. "Parallel and Serial Processes in Number-to-Quantity Conversion." *Cognition* 204(2020): 1–17.

Drach, Kostiantyn, and Richard Evan Schwartz. "A Hyperbolic View of the Seven Circles Theorem." *The Mathematical Intelligencer* 42.2(2020): 61–65.

Dragović, Vladimir, and Irina Goryuchkina. "Polygons of Petrović and Fine, Algebraic ODEs, and Contemporary Mathematics." *Archive for History of Exact Sciences* 74.6(2020): 523–564.

Dunham, William. "Odd Perfect Numbers: A Triptych." *The Mathematical Intelligencer* 42.1(2020): 42–46.

Durst, Susan, and Scott R. Kaschner. "Logical Misconceptions with Conditional Statements in Early Undergraduate Mathematics." *PRIMUS* 30.4(2020): 367–378.

Endress, Ansgar D., Lauren K. Slone, and Scott P. Johnson. "Statistical Learning and Memory." *Cognition* 204(2020): 1–9.

Enea, Maria Rosaria, and Giovanni Ferraro. "Analytic and Arithmetic Methods in Liouville's Identities." *Historia Mathematica* 53(2020): 48–70.

Ewing, E. Thomas, Steven E. Rigdon, and Ronald D. Fricker Jr. "Understanding COVID-19 in 2020 through the Lens of the 1918 'Spanish Flu' Epidemic." *Chance* 33.3(2020): 4–21.

Fan, Zhao. "Hobson's Conception of Definable Numbers." *History and Philosophy of Logic* 41.2(2020): 128–139.

Fang, Zhihui, and Suzanne Chapman. "Disciplinary Literacy in Mathematics: One Mathematician's Reading Practices." *The Journal of Mathematical Behavior* 59(2020): 1–15.

Ferraro, Giovanni. "Euler and the Structure of Mathematics." *Historia Mathematica* 50(2020): 2–24.

Friendly, Michael, and Howard Wainer. "Galton's Gleam: Visual Thinking and Graphic Discoveries." *Significance* 17.3(2020): 28–33.

Gabel, Mika, and Tommy Dreyfus. "Analyzing Proof Teaching at the Tertiary Level Using Perelman's New Rhetoric." *For the Learning of Mathematics* 40.2(2020): 15–19.

Gelman, Andrew, and Alexey Guzey. "Statistics as Squid Ink: How Prominent Researchers Can Get Away with Misrepresenting Data." *Chance* 33.2(2020): 25–27.

Gentili, Graziano, Luisa Simonutti, and Daniele C. Struppa. "The Mathematics of the Astrolabe and Its History." *Journal of Humanistic Mathematics* 10.1(2020): 101–144.

Ginoux, Jean-Marc, and Jean-Claude Golvin. "Perimeter Determination of the Eight-Centered Oval." *The Mathematical Intelligencer* 42.2(2020): 20–29.

Goldstein, Harvey. "Living by the Evidence." *Significance* 17.1(2020): 38–40.

Gouet, Camilo, Salvador Carvajal, Justin Halberda, and Marcela Peña. "Training Nonsymbolic Proportional Reasoning in Children and Its Effects on Their Symbolic Math Abilities." *Cognition* 197(2020): 1–13.

Grohe, Martin, and Pascal Schweitzer. "The Graph Isomorphism Problem." *Communications of the ACM* 63.11(2020): 128–134.

Grove, Michael J., and Chris Good. "Approaches to Feedback in the Mathematical Sciences: Just What Do Students Really Think?" *Teaching Mathematics and its Applications* 39.3(2020): 160–183.

Gustar, Andrew. "The Laws of Musical Fame and Obscurity." *Significance* 17.5(2020): 14–17.

Hall, Rachel Wells. "Math for Drummers." *UMAP Journal* 41.1(2020): 37–50.

Hall, Rachel Wells. "Math for Poets." *UMAP Journal* 41.2(2020): 145–176.

Hankeln, Corinna. "Mathematical Modeling in Germany and France: A Comparison of Students' Modeling Processes." *Educational Studies in Mathematics* 103(2020): 209–229.

Hanusch, Sarah. "Summative Portfolios in Undergraduate Mathematics Courses." *PRIMUS* 30.3(2020): 274–284.

Haridis, Alexandros. "Structure from Appearance: Topology with Shapes, without Points." *Journal of Mathematics and the Arts* 14.3(2020): 199–238.

Harkness, Timandra. "John Graunt at 400: Fighting Disease with Numbers." *Significance* 17.4(2020): 22–25.

Henle, Jim, and Craig Kasper. "A Flowering of Mathematical Art." *The Mathematical Intelligencer* 42.1(2020): 36–40.

Herbert, Annie, Gareth Griffith, Gibran Hemani, and Luisa Zuccolo. "The Spectre of Berkson's Paradox: Collider Bias in Covid-19 Research." *Significance* 17.4(2020): 6–7.

Hirsch, David, and Katherine A. Seaton. "The Polycons: The Sphericon (or Tetracon) Has Found Its Family." *Journal of Mathematics and the Arts* 14.4(2020): 345–359.

Hoa, Lê Tuấn. "The Development of Mathematical Research in Vietnam at a Glance." *The Mathematical Intelligencer* 42.4(2020): 50–58.

Hobbs, Stephen L., and Susan D. Nickerson. "Optimizing Disaster Relief from a Navy Aircraft Carrier." *UMAP Journal* 41.2(2020): 101–120.

Holmes, Lisa, Katie Edwards, Andy Moss, Simon Tollington, Anna Fowler, David Hughes, and Maria Sudell. "Statistics at the Zoo." *Significance* 17.5(2020): 26–29.

Huber, Mark. "A Probabilistic Approach to the Fibonacci Sequence." *The Mathematical Intelligencer* 42.3(2020): 29–33.

Ibrahim, Maria. "Primum non nocere (First, Do No Harm)." *Significance* 17.6(2020): 24–26.

Iliev, Iliyan R., Xin Huang, and Yulia R. Gel. "Speaking Out or Speaking In. Changes in Political Rhetoric over Time." *Significance* 17.5(2020): 22–25.

Ingram, Jenni. "Epistemic Management in Mathematics Classroom Interactions: Student Claims of Not Knowing or Not Understanding." *The Journal of Mathematical Behavior* 58(2020): 1–13.

Jackson, James. "Statistics and the 'Little Grey Cells'." *Significance* 17.5(2020): 10–11.

Jaëck, Frédéric. "What Were the Genuine Banach Spaces in 1922? Reflection on Axiomatisation and Progression of the Mathematical Thought." *Archive for History of Exact Sciences* 74.2(2020): 109–129.

Jakobsen, David. "A.N. Prior and 'The Nature of Logic'." *History and Philosophy of Logic* 41.1(2020): 71–81.

Jebeile, Julie, and Michel Crucifix. "Multi-Model Ensembles in Climate Science: Mathematical Structures and Expert Judgements." *Studies in History and Philosophy of Science, Part A* 83(2020): 44–52.

Ji, Lizhen, and Chang Wang. "Poincaré's Stated Motivations for Topology." *Archive for History of Exact Sciences* 74.4(2020): 381–400.

Joseph, Toby. "An Alternative Proof of Euler's Rotation Theorem." *The Mathematical Intelligencer* 42.4(2020): 44–49.

Kanellopoulos, Giorgos, Dimitrios Razis, and Ko van der Weele. "The Persian Immortals: A Classical Case of Self-Organization." *American Journal of Physics* 88.4(2020): 263–268.

Katz, Mikhail G. "Mathematical Conquerors, Unguru Polarity, and the Task of History." *Journal of Humanistic Mathematics* 10.1(2020): 475–515.

Katz, Sophia. "Structure and Numbers: Shao Yong on the Order of Reality." *Studies in History and Philosophy of Science, Part A* 81(2020): 16–23.

Khamjane, Aziz, Rachid Benslimane, and Zouhair Ouazene. "Method of Construction of Decagonal Self-Similar Patterns." *Nexus Network Journal* 22.2(2020): 507–520.

Khorami, Mehdi. "Space Harmony: A Knot Theory Perspective on the Work of Rudolf Laban." *Journal of Mathematics and the Arts* 14.3(2020): 239–257.

Kim, Joongol. "The Adverbial Theory of Numbers: Some Clarifications." *Synthese* 197(2020): 3981–4000.

Klarreich, Erica. "'Amazing' Math Bridge Extended beyond Fermat's Last Theorem." *Quanta Magazine* April 6, 2020. https://www.quantamagazine.org/amazing-math-bridge -extended-beyond-fermats-last-theorem-20200406/.

Klarreich, Erica. "Multiplication Hits the Speed Limit." *Communications of the ACM* 63.1(2020): 11–13.

Klymchuk, Sergiy, and Kerri Spooner. "University Students' Preferences for Application Problems and Pure Mathematics Questions." *Teaching Mathematics and Its Applications* 39.1(2020): 29–37.

Kontorovich, Igor'. "Theorems or Procedures? Exploring Undergraduates' Methods to Solve Routine Problems in Linear Algebra." *Mathematics Education Research Journal* 32.4(2020): 589–605.

Kuper, Emily, and Marilyn Carlson. "Foundational Ways of Thinking for Understanding the Idea of Logarithm." *The Journal of Mathematical Behavior* 57(2020): 1–18.

Kuperberg, Greg. "Using the arXiv." *Notices of the American Mathematical Society* 67.2(2020): 184–187.

Langkjær-Bain, Robert. "Not Buying It!" *Significance* 17.4(2020): 16–21.

Laraudogoitia, Jon Pérez. "Some Surprising Instabilities in Idealized Dynamical Systems." *Synthese* 197.3(2020): 3007–3026.

Lastra, Alberto, and Manuel De Miguel. "Geometry of Curves and Surfaces in Contemporary Chair Design." *Nexus Network Journal* 22.3(2020): 643–657.

Laudau, Elizabeth. "How the Mathematical Conundrum Called the 'Knapsack Problem' Is All around Us." *Smithsonian Magazine Online* March 9, 2020. https://www.smithsonianmag .com/science-nature/why-knapsack-problem-all-around-us-180974333/.

Leng, Mary. "An '*i*' for an *i*, a Truth for a Truth." *Philosophia Mathematica* 28.3(2020): 347–359.

Lew, Kristen, and Juan Pablo Mejía Ramos. "Linguistic Conventions of Mathematical Proof Writing across Pedagogical Contexts." *Educational Studies in Mathematics* 103(2020): 43–62.

Lockwood, Elise, John S. Caughman, and Keith Weber. "An Essay on Proof, Conviction, and Explanation: Multiple Representation Systems in Combinatorics." *Educational Studies in Mathematics* 103.3(2020): 173–189.

Lockwood, Elise, Nicholas H. Wasserman, and Erik S. Tillema. "A Case for Combinatorics: A Research Commentary." *The Journal of Mathematical Behavior* 59.1(2020): 1–15.

Lorenat, Jemma. "'Actual Accomplishments in This World': The Other Students of Charlotte Angas Scott." *The Mathematical Intelligencer* 42.1(2020): 56–65.

Lorenat, Jemma. "Drawing on the Imagination: The Limits of Illustrated Figures in Nineteenth-Century Geometry." *Studies in History and Philosophy of Science, Part A* 82(2020): 75–87.

Louis, Winnifred, and Cassandra Chapman. "The Seven Deadly Sins of Statistical Misinterpretation and How to Avoid Them." *The Conversation* March 28, 2017. https://theconversation .com/the-seven-deadly-sins-of-statistical-misinterpretation-and-how-to-avoid-them -74306.

MacPherson-Krutsky, Carson. "3 Questions to Ask Yourself Next Time You See a Graph, Chart or Map." *The Conversation* July 24, 2020. https://theconversation.com/3-questions -to-ask-yourself-next-time-you-see-a-graph-chart-or-map-141348

Makovicky, Emil. "Byzantine Floor Patterns of Interlocking Twisted Cable Loops." *Nexus Network Journal* 22.2(2020): 449–469.

Marchetti, Elena, and Luisa Rossi Costa. "The House of the White Man: A Mathematical Description." *Nexus Network Journal* 22.3(2020): 631–641.

Marotta, Anna, Ursula Zich, and Martino Pavignano. "Fortification Design and Geometry in the Papers of Gaspare Beretta." *Nexus Network Journal* 22.1(2020): 169–190.

Masson, Nicolas, Michael Andres, Marie Alsamour, Zoé Bollen, and Mauro Pesenti. "Spatial Biases in Mental Arithmetic Are Independent of Reading/Writing Habits: Evidence from French and Arabic Speakers." *Cognition* 200(2020): 1–10.

Mazur, Barry. "Math in the Time of Plague." *The Mathematical Intelligencer* 42.4(2020): 1–6.

McBride, Rebekah. "The Stories behind the Data." *Significance* 17.2(2020): 10–11.

Meyer, Michael, and Susanne Schnell. "What Counts as a 'Good' Argument in School? How Teachers Grade Students' Mathematical Arguments." *Educational Studies in Mathematics* 105(2020): 35–51.

Molenaar, Jaap.* "Mathematics, the Science of My Life." *Nieuw Archief voor Wiskunde* 5/21.1(2020): 38–45.

Moradzadeh, Sam, and Ahad Nejad Ebrahimi. "Islamic Geometric Patterns in Higher Dimensions." *Nexus Network Journal* 22.3(2020): 777–798.

Moreno, Fernando Díaz. "Hand Drawing in the Definition of the First Digital Curves." *Nexus Network Journal* 22.3(2020): 755–775.

Nolan, Deborah, and Sara Stoudt. "Reading to Write." *Significance* 17.6(2020): 34–37.

Northcott, Robert. "Big Data and Prediction: Four Case Studies." *Studies in History and Philosophy of Science, Part A* 81(2020): 96–104.

Nothaft, C. Philipp E. "Medieval Europe's Satanic Ciphers: On the Genesis of a Modern Myth." *British Journal for the History of Mathematics* 35.2(2020): 107–136.

Nutting, Eileen S. "Benacerraf, Field, and the Agreement of Mathematicians." *Synthese* 197(2020): 2095–2110.

Olsen, Joe, Kristen Lew, and Keith Weber. "Metaphors for Learning and Doing Mathematics in Advanced Mathematics Lectures." *Educational Studies in Mathematics* 105(2020): 1–17.

Ostwald, Michael J., and Michael J. Dawes. "The Spatio-Visual Geometry of the *Hollyhock House*: A Mathematical Analysis of the 'Wright Space' Using Isovist Fields." *Nexus Network Journal* 22.1(2020): 211–228.

Oswald, Nicola, and Jörn Steuding. "A Hidden Orthogonal Latin Square in a Work of Euler from 1770." *The Mathematical Intelligencer* 42.4(2020): 23–29.

Papayannopoulos, Philippos. "Computing and Modelling: Analog vs. Analogue." *Studies in History and Philosophy of Science, Part A* 83(2020): 103–120.

Pinto, Alon, and Ronnie Karsenty. "Norms of Proof in Different Pedagogical Contexts." *For the Learning of Mathematics* 40.1(2020): 22–27.

Pollak, Henry. "Down with Base-10 Logarithms!" *UMAP Journal* 41.2(2020): 93–100.

Ramos-Jaime, Cristina, and José Sánchez-Sánchez. "Hyperboloid Modules for Deployable Structures." *Nexus Network Journal* 22.2(2020): 309–328.

Raper, Simon. "The End of Determinism." *Significance* 17.6(2020): 14–17.

Rasmussen, Chris, Naneh Apkarian, Michal Tabach, and Tommy Dreyfus. "Ways in Which Engaging with Someone Else's Reasoning Is Productive." *The Journal of Mathematical Behavior* 58(2020): 1–23.

Richeson, David. "Communicating Mathematics Using Social Media." *Notices of the American Mathematical Society* 67.2(2020): 182–184.

Ródenas-López, Manuel Alejandro, Martino Peña Fernández-Serrano, Pedro Miguel Jiménez-Vicario, Pedro García Martínez, and Adolfo Pérez Egea. "Geometric Evaluation of Deployable Structures Using Parametric Modelling." *Nexus Network Journal* 22.1(2020): 247–270.

Roeper, Peter. "Reflections on Frege's Theory of Real Numbers." *Philosophia Mathematica* 28.2(2020): 236–257.

Roselló, Vicenta Calvo, Esther Capilla Tamborero, and Juan Carlos Navarro Fajardo. "Oval Domes: The Case of the *Basílica de la Virgen de los Desamparados* of Valencia." *Nexus Network Journal* 22.2(2020): 393–409.

Rosso, Riccardo. "Probability and Exams: The Work of Antonio Bordoni." *Historia Mathematica* 53(2020): 33–47.

Rozhkovskaya, Natasha. "Mathematical Commentary on Le Corbusier's Modulor." *Nexus Network Journal* 22.2(2020): 411–428.

Rubin, Mark. "That's Not a Two-Sided Test! It's Two One-Sided Tests!" *Significance* 17.3(2020): 38–41.

Sabetfard, Mojtaba, and Hadi Nadimi. "Generating Square Kufic Patterns Using Cellular Automata." *Nexus Network Journal* 22.2(2020): 275–290.

Sacchetti, Andrea. "Francesco Carlini: Kepler's Equation and the Asymptotic Solution to Singular Differential Equations." *Historia Mathematica* 53(2020): 1–32.

Schuckers, Michael, and Sarah Campbell. "Telling Statistical Stories." *Significance* 17.6(2020): 30–33.

Sella, Francesco, and Daniela Lucangeli. "The Knowledge of the Preceding Number Reveals a Mature Understanding of the Number Sequence." *Cognition* 194(2020): 1–14.

Shmahalo, Olena. "How Gödel's Proof Works." *Quanta Magazine* July 14, 2020. https://www.quantamagazine.org/how-godels-incompleteness-theorems-work-20200714/.

Skovsmose, Ole. "Three Narratives about Mathematics Education." *For the Learning of Mathematics* 40.1(2020): 47–51.

Small, Hugh. "Nightingale's Overlooked Scutari Statistics." *Significance* 17.6(2020): 28–33.

Soto, Hortensia. "Teaching on Purpose and with a Purpose: The Scarecrow, Lion & Tin Woodman." *PRIMUS* 30.7(2020): 827–836.

Spindler, Richard. "Aligning Modeling Projects with Bloom's Taxonomy." *PRIMUS* 30.5(2020): 601–616.

Stenhouse, Brigitte. "Mary Somerville's Early Contributions to the Circulation of Differential Calculus." *Historia Mathematica* 51(2020): 1–25.

Sullivan, Peter, Janette Bobis, Ann Downton, Maggie Feng, Sally Hughes, Sharyn Livy, Melody McCormick, and James Russo. "Threats and Opportunities in Remote Learning of Mathematics: Implication for the Return to the Classroom." *Mathematics Education Research Journal* 32.3(2020): 551–559.

Thomas, Robert S. D. "Why Our Hand is Not the Whole Deck: Embrace, Acceptance, or Use of Limitations." *Journal of Humanistic Mathematics* 10.1(2020): 267–294.

Unger, J. Marshall. "On the Acceptance of Trigonometry in *Wasan*: Evidence from a Text of Aida Yasuaki." *Historia Mathematica* 52(2020): 51–65.

Vaccaro, Maria Alessandra. "Historical Origins of the Nine-Point Conic. The Contribution of Eugenio Beltrami." *Historia Mathematica* 51(2020): 26–48.

Wagner, David, Arthur Bakker, Tamsin Meaney, Vilma Mesa, Susanne Prediger, and Wim Van Dooren. "What Can We Do against Racism in Mathematics Education Research?" *Educational Studies in Mathematics* 104(2020): 299–311.

Walkoe, Janet, and Mariana Levin. "Seeds of Algebraic Thinking: Towards a Research Agenda." *For the Learning of Mathematics* 40.2(2020): 27–31.

Wares, Arsalan. "Mathematical Art and Artistic Mathematics." *International Journal of Mathematical Education in Science and Technology* 51.1(2020): 152–156.

Weber, Keith, Juan Pablo Mejía-Ramos, Timothy Fukawa-Connelly, and Nicholas Wasserman. "Connecting the Learning of Advanced Mathematics with the Teaching of

Secondary Mathematics: Inverse Functions, Domain Restrictions, and the Arcsine Function." *The Journal of Mathematical Behavior* 57(2020): 1–21.

Wheeler, Nicole. "Computer Says . . ." *Significance* 17.3(2020): 34–37.

Wilhelm, Isaac. "A Statistical Analysis of Luck." *Synthese* 197(2020): 867–885.

Wilmott, Paul, and David Orrell. "No Laws, Only Toys." *Wilmott Magazine* 99(2019): 20–29.

Woods, Christian, and Keith Weber. "The Relationship between Mathematicians' Pedagogical Goals, Orientations, and Common Teaching Practices in Advanced Mathematics." *The Journal of Mathematical Behavior* 59(2020): 1–17.

Zhu, Yiwen. "How Do We Understand Mathematical Practices in Non-Mathematical Fields? Reflections Inspired by Cases from 12th and 13th Century China." *Historia Mathematica* 52(2020): 1–25.

Notable Journal Issues

The following journal issues are fully or partly dedicated to the specified topic—or contain symposia on the respective theme.

"Probabilistic Knowledge in Action." *Analysis* 80.2(2020).

"Mathematics, Networks and Practices around Early Modern Scotland." *British Journal for the History of Mathematics* 35.1(2020).

"A New Generation of Statisticians Tackles Data Privacy." *Chance* 33.4(2020).

"Affect in Mathematical Problem Posing." *Educational Studies in Mathematics* 105.3(2020).

"The Foundations of Software Science and Computational Structures." *Foundations of Science* 25.4(2020).

"Studies in Post-Medieval Logic." *History and Philosophy of Logic* 41.4(2020).

"Creativity in Mathematics." *Journal of Humanistic Mathematics* 10.2(2020).

Artists' Viewpoints [on Mathematics]. *Journal of Mathematics and the Arts* 14.1–2(2020).

"Spontaneous Mathematical Focusing Tendencies in Mathematical Development." *Mathematical Thinking and Learning* 22.4(2020).

"Inclusive Mathematics Education." *Mathematics Education Research Journal* 32.1(2020).

"The Relation between Mathematics Achievement and Spatial Reasoning." *Mathematics Education Research Journal* 32.2(2020).

"Mathematics Teaching and Learning in Malaysia." *Mathematics Enthusiast* 17.1(2020).

"Supporting Mathematics Teacher Educators' Knowledge and Practices for Teaching Content to Prospective (Grades K-8) Teachers." *Mathematics Enthusiast* 17.2–3(2020).

"Complex Vaulted Systems: Geometry and Architecture from Design to Construction." *Nexus Network Journal* 22.4(2020).

"Foundations of Mathematical Structuralism." *Philosophia Mathematica* 28.3(2020).

"Leading a Department in the Mathematical Sciences." *PRIMUS [Problems, Resources, and Issues in Mathematics Undergraduate Studies]* 30.6(2020).

"Implementing Mastery Grading in the Undergraduate Mathematics Classroom." *PRIMUS [Problems, Resources, and Issues in Mathematics Undergraduate Studies]* 30.8–10(2020).

"Florence Nightingale." *Significance* 17.2(2020).

"Foundations of Mathematics." *Synthese* 197.2(2020).

"Mathematical Cognition and Enculturation." *Synthese* 197.9(2020).

"2020 Interdisciplinary Contest in Modeling." *UMAP Journal* 41.3(2020).

"Mathematical Word Problem Solving: Psychological and Educational Perspectives." *Zentralblatt für Didaktik der Mathematik* 52.1(2020).

"Expertise in Developing Students' Expertise in Mathematics." *Zentralblatt für Didaktik der Mathematik* 52.2(2020).

"Numeracy and Vulnerability in Adult Life." *Zentralblatt für Didaktik der Mathematik* 52.3(2020).

"Research on Early Childhood Mathematics Teaching and Learning." *Zentralblatt für Didaktik der Mathematik* 52.4(2020).

"Online Mathematics Education and e-Learning." *Zentralblatt für Didaktik der Mathematik* 52.5(2020).

"The Role of Mathematicians' Practice in Mathematics Education Research." *Zentralblatt für Didaktik der Mathematik* 52.6(2020).

"Teaching with Digital Technology." *Zentralblatt für Didaktik der Mathematik* 52.7(2020).

Acknowledgments

As I mentioned in the introduction, this series of annual anthologies of writings on mathematics faces an uncertain future. I initiated it and I edited it for 12 years; I thank all the contributors to all the volumes for writing the pieces, as well as the original publishers for granting copyright permissions.

At Princeton University Press, thank you to Vickie Kearn (now retired) for accepting my book series proposal in 2009 and for guiding me with her expert advice throughout the preparation of the first 10 volumes. Also thanks to Susannah Shoemaker, who took over Vickie's job two years ago, and to her assistant Kristen Hop. Nathan Carr oversaw the production of each volume in the series, with exquisite attention to details and deadlines—thank you, Nathan! Many thanks to Paula Bérard for copyediting 11 of the volumes, including this one. Also thanks to the readers and reviewers who, indirectly but unfailingly, helped us every year to reach the final table of contents; almost all remained anonymous to me, except when I discovered by accident who they were.

Thank you to the libraries at Cornell University and Syracuse University; without their borrowing services and electronic access to journals, I could not have done the bibliographic research required by this enterprise.

I started the series while I was a graduate student at Cornell University. My gratitude for the opportunities offered to me during those years goes foremost to the memory of the late David Henderson, my main academic advisor—but also to Steven Strogatz and John Sipple (my other advisors), and to Maria Terrell, who assigned me teaching assistantships that kept me afloat during a time of extreme financial hardship.

In the department of mathematics at Syracuse University, thanks to Leonid Kovalev, Graham Leuschke, and Jeffrey Meyer for giving me plenty of courses to teach and for scheduling my teaching conveniently, so that I have time for other endeavors. This allows me to scale up skills I learned before I got this teaching job.

In the family, thanks to my daughter Ioana for growing into a responsible adult after a difficult childhood; thanks to my sons Leo and Ray for their laughter and playfulness; and thanks especially to my wife Fangfang for her love, her humor, her cooking, her gardening, . . . !

Credits